ERGEBNISSE DER MATHEMATIK
UND IHRER GRENZGEBIETE
UNTER MITWIRKUNG DER SCHRIFTLEITUNG DES
„ZENTRALBLATT FÜR MATHEMATIK“

HERAUSGEGEBEN VON
L. V. AHLFORS · R. BAER · F. L. BAUER · R. COURANT · A. DOLD
J. L. DOOB · S. EILENBERG · P. R. HALMOS · M. KNESER
T. NAKAYAMA · H. RADEMACHER · F. K. SCHMIDT
B. SEGRE · E. SPERNER

NEUE FOLGE · HEFT 27

# NORM IDEALS OF COMPLETELY CONTINUOUS OPERATORS

BY

ROBERT SCHATTEN

SPRINGER-VERLAG
BERLIN · GÖTTINGEN · HEIDELBERG
1960

ERGEBNISSE DER MATHEMATIK UND IHRER GRENZGEBIETE

UNTER MITWIRKUNG DER SCHRIFTLEITUNG DES „ZENTRALBLATT FÜR MATHEMATIK“



HERAUSGEGEBEN VON

L. V. AHLFORS · R. BAER · F. L. BAUER · R. COURANT · A. DOLD
J. L. DOOB · S. EILENBERG · P. R. HALMOS · M. KNESER
T. NAKAYAMA · H. RADEMACHER · F. K. SCHMIDT
B. SEGRE · E. SPERNER



NEUE FOLGE · HEFT 27


REIHE:

# REELLE FUNKTIONEN

BESORGT
VON

P. R. HALMOS


SPRINGER-VERLAG
BERLIN · GÖTTINGEN · HEIDELBERG
1960

# NORM IDEALS OF COMPLETELY CONTINUOUS OPERATORS

BY

ROBERT SCHATTEN

SPRINGER-VERLAG
BERLIN · GÖTTINGEN · HEIDELBERG
1960

ISBN-13: 978-3-642-87654-7 e-ISBN-13: 978-3-642-87652-3

DOI: 10.1007/978-3-642-87652-3

# Preface

Completely continuous operators on a Hilbert space or even on a Banach space have received considerable attention in the last fifty years. Their study was usually confined to special completely continuous operators or to the discovery of properties common to all of them (for instance, that every such operator admits a proper invariant subspace).

On the other hand, interest in *spaces* of completely continuous operators is comparatively new. Some results of this type may be found implicit in the early work of E. SCHMIDT. Other results are "generally known" and cannot be found explicitly in print. One of the interesting and relatively new results states that modulo the language of BANACH (that is, up to equivalence) the space of all operators on a Hilbert space $\mathfrak{H}$ is the second conjugate of the space of all completely continuous operators on $\mathfrak{H}$.

The study of spaces of completely continuous operators on a perfectly general Banach space involves many difficulties. Some stem, for instance, from the unsolved problem whether a completely continuous operator on a perfectly general Banach space is always approximable in bound by operators of finite rank. The answer is affirmative in all the special Banach spaces considered. An affirmative answer to the above problem is the ultimate desideratum — it would simplify the theory considerably. A negative answer, however, would be equally interesting (although for us not so useful), since it would settle negatively the open "basis problem". (The range of a completely continuous operator $A$ may be always assumed to lie in a separable Banach space; the last must not have a basis whenever $A$ is not approximable in bound by operators of finite rank.)

However, at this stage of the game it is possible — and quite timely — to present a reasonable theory of spaces of completely continuous operators on a Hilbert space. That is precisely what we plan to accomplish. For the sake of simplicity we shall ignore the fact that fragments of the results which we are about to present may be carried over in a straightforward manner to completely continuous operators on an arbitrary Banach space.

The setting for our discussion is a complex Hilbert space $\mathfrak{H}$; we denote by $\mathfrak{A}$ the algebra of all operators on $\mathfrak{H}$. The Banach spaces considered herein are made up of completely continuous operators on $\mathfrak{H}$, and the norm $\alpha(A)$ of an operator $A$ is in general different from its bound $\|A\|$. It turns out that the most important spaces are also the interesting ones; every one of such spaces $\mathscr{B}$ is also a two-sided ideal in the algebra $\mathfrak{A}$, and the norm on $\mathscr{B}$ satisfies the extra condition

$$\alpha(XAY) \leqq \|X\| \, \|Y\| \, \alpha(A)$$

for $A \in \mathscr{B}$ and $X$, $Y$ in $\mathfrak{A}$. Motivated by the above considerations we are led to the study of *norm-ideals* of completely continuous operators on a Hilbert space.

For some reason, spaces of completely continuous operators have received little attention in the literature, too little when one considers the rapid and significant progress of the operator theory in the last few decades. It is hoped therefore that this not too pretentious presentation will induce the interested reader to further investigations in this promising field.

Those familiar with the contributions of JOHN VON NEUMANN will no doubt recognize here the influence of some of his work. It is to his memory that this monograph is dedicated.

Thanks are due to the National Science Foundation for financial support during the academic year 1957—1958 and to The University of Kansas Research Fund for a grant in 1958—1959.

R. S.

Lawrence, Kansas

January 1960

# Contents

# Preliminaries and notation

## 1. Prerequisites

The prerequisites for a thorough understanding of this essentially self-contained exposition are quite modest. Rather than to write a small textbook on these, or go to the other extreme and say that these are well known and may be easily found in the literature, some sort of a middle road seems to offer a more satisfactory solution. We state a sufficient number of references and merely redefine a few notions, some of which, though not overemphasized in standard texts, play an important part in our discussion. We also reconstruct the proof of the "polar decomposition" for operators which indeed plays a central role in this exposition.

The first five chapters of STONE's [1] treatise or the first 25 pages of NAGY's [1] concise exposition include more than the necessary preliminaries required of Hilbert space geometry and of the theory of operators on those spaces. The same may be said for the first two chapters of the more recent exposition by HALMOS [2]; the terminology and notation used therein — especially when it involves infinite sums — is closest to the one adopted here. Of course, the book on finite-dimensional vector spaces by HALMOS [3] as well as the one by HAMBURGER and GRIMSHAW [1] may be found to be very helpful in the present study inasmuch as some of the properties of linear transformations on a finite-dimensional space carry over in a straightforward manner to completely continuous operators on a Hilbert space.

The little information required of Banach spaces may be found in the famous treatise of BANACH [1] or even in the very readable and brief presentation due to KOLMOGOROFF and FOMIN [1]. For the elementary few definitions involving Banach algebras we refer the reader to a few pages in the book by LOOMIS [1, §§ 11, 18, 26] or to the one by HILLE and PHILLIPS [1, §§ 1.4 and 4.6].

Some of the ideas involved in this study may be already found in SCHATTEN [6]. The last book is an outgrowth of a series of papers written by this author either alone [1, 2, 3, 4, 5, 7] or in collaboration with JOHN VON NEUMANN [1, 2]. The reader will do well consulting these occasionally. Much has been added in this connection since their appearance. We merely mention two papers by RUSTON [1, 2] and an extensive monograph due to GROTHENDIECK [1].

## 2. Hilbert space

The setting for most of our discussion is a fixed complex Hilbert space $\mathfrak{H}$, that is, a complex vector space with an inner product for which the resulting metric space is complete. Completeness is an essential part of the structure of $\mathfrak{H}$, although any linear space with an inner product $\mathfrak{R}$ can be always imbedded in a Hilbert space $\mathfrak{H}$. We may add that this imbedding can be accomplished in one and only one manner if $\mathfrak{R}$ is to be dense in $\mathfrak{H}$; the last is then the ordinary metric (CANTOR-MERAY) completion of $\mathfrak{R}$.

A basis in $\mathfrak{H}$ is by definition a maximal orthonormal family of vectors $\{\varphi_i\}$. It follows from Zorn's lemma (which is equivalent to Zermelo's axiom of choice or to the theorem that for any set there exists a well-ordering) that in any Hilbert space there is a basis. The cardinal number corresponding to any two bases in $\mathfrak{H}$ is always the same and defines the dimension of the space (see LÖWIG [1]). A finite-dimensional space is often referred to as a unitary space; we shall denote such by $\mathfrak{H}_n$ if its dimension is $n$. On the other hand a space of dimension $\aleph_0$ is said to define a "true" Hilbert space. Clearly, a Hilbert space is separable, that is, contains a denumerable dense subset if and only if its dimension is $\leqq \aleph_0$. For any cardinal number $m$ there is a Hilbert space of dimension $m$.

Let $(\mathfrak{X}, S, \nu)$ be a given measure space. We denote by $\mathfrak{L}^2$ the set of all complex valued measurable functions $f(x)$ on $\mathfrak{X}$ for which $|f|^2$ is integrable; two functions are considered identical if and only if they differ on a set of measure zero. Strictly speaking we are considering the quotient space $\mathfrak{L}^2/\mathfrak{N}$ where $\mathfrak{N}$ is the set of all measurable functions equal to zero almost everywhere. There, the linear operations are the usual ones in function spaces and the inner product is given by

$$(f, g) = \int f(x)\, \overline{g(x)}\, d\nu \,.$$

Completeness of the above space is the content of a theorem due to F. RIESZ and E. FISCHER.

In particular, if $\mathfrak{X}$ is a set whose cardinal is $m$, $S$ is a $\sigma$-ring of "measurable" sets containing among others all the finite subsets of $\mathfrak{X}$, and the measure $\nu$ of a measurable set is defined to be the number of points it contains, then $\mathfrak{L}^2$ consists of all complex-valued functions on $\mathfrak{X}$, vanishing everywhere except at most at a denumerable number of points and for which the series of squares of the absolute of those non-vanishing values converges; the dimension of $\mathfrak{L}^2$ is obviously $m$ (compare HALMOS [1]).

Choosing for $\mathfrak{X}$ the interval $0 \leqq x \leqq 1$ (square $0 \leqq x, y \leqq 1$), for $S$ the class of Lebesgue measurable sets in that interval (square), for $\nu$ the linear (planar) Lebesgue measure, the resulting special $\mathfrak{L}^2$ will be denoted by $L^2$ ($\boldsymbol{L}^2$).

## 3. Summability

Since we are also dealing with non-separable spaces, it is convenient to make use of the notion of summability. A family $\{f_j\}$ of vectors (where $j$ varies over some set of indices) is said to be summable with sum $f$, in symbol $\Sigma_j f_j = f$, if for every $\varepsilon > 0$ there exists a finite set $J_0 = J_0(\varepsilon)$ of indices, such that $\|f - \Sigma_{j \in J} f_j\| < \varepsilon$ for every finite set of indices $J \supset J_0$. This definition is clearly independent of any ordering of the vectors in $\{f_j\}$. Any finite family is summable with a sum equal to the ordinary vector sum. If $\{f_j\}$ is summable, then $f_j \neq 0$ for at most a countable number of indices. The summability of a *sequence* $\{f_i\}$ of orthogonal vectors (with sum $f$) is equivalent to ordinary — necessarily unconditional — convergence of the series $f_1 + f_2 + \cdots$ (and $f$ is its sum). An orthogonal family of vectors $\{f_j\}$ is summable if and only if the corresponding family $\{\|f_j\|^2\}$ of non-negative numbers is summable; if $f = \Sigma_j f_j$ then $\|f\|^2 = \Sigma_j \|f_j\|^2$. We shall also write $\Sigma_j a_j < +\infty$ to express that the family $\{a_j\}$ of non-negative numbers is summable.

The following conditions on an orthonormal family of vectors $\{\varphi_j\}$ are equivalent:

1. $\{\varphi_j\}$ is a basis in $\mathfrak{H}$ (that is, $\{\varphi_j\}$ is "complete").
2. $(f, \varphi_j) = 0$ for all $j$, implies $f = 0$.
3. For every $f \in \mathfrak{H}$ we have $f = \Sigma_j (f, \varphi_j)\, \varphi_j$ (Fourier expansion).
4. For every pair $f, g$ in $\mathfrak{H}$ we have
$$\Sigma_j (f, \varphi_j)(\varphi_j, g) = (f, g) \qquad \text{(Parseval's identity)}.$$
5. For $f \in \mathfrak{H}$, we have $\|f\|^2 = \Sigma_j |(f, \varphi_j)|^2$.

## 4. Subspaces

A linear manifold is a subset $\mathfrak{M}$ of $\mathfrak{H}$ which contains all the linear combinations of any two vectors in $\mathfrak{M}$. A subspace is a closed linear manifold. The span of a subset $\mathfrak{S}$ of $\mathfrak{H}$, in symbol $\vee\, \mathfrak{S}$, is defined as the least subspace containing $\mathfrak{S}$.

If $\mathfrak{M}_1$ and $\mathfrak{M}_2$ are two subspaces, then the set of all vectors in the form $f_1 + f_2$ with $f_1 \in \mathfrak{M}_1$ and $f_2 \in \mathfrak{M}_2$ may not be closed. If, however, $\{\mathfrak{M}_j\}$ is a family of pairwise orthogonal subspaces, then the set of all vectors of the form $\Sigma_j f_j$ with $f_j \in \mathfrak{M}_j$ for all $j$, is also a subspace. The last is the least subspace containing all the $\mathfrak{M}_j$ and will be denoted by $\Sigma_j \oplus \mathfrak{M}_j$ and in the case of a sequence of subspaces also $\mathfrak{M}_1 \oplus \mathfrak{M}_2 \oplus \cdots$.

If $\mathfrak{M}$ and $\mathfrak{N}$ are two subspaces and $\mathfrak{N} \subset \mathfrak{M}$, then $\mathfrak{M} \ominus \mathfrak{N}$ is the subspace consisting of all the vectors in $\mathfrak{M}$ which are orthogonal to $\mathfrak{N}$. We have, $\mathfrak{M} = \mathfrak{N} \oplus (\mathfrak{M} \ominus \mathfrak{N})$.

## 5. Operators

A bounded linear transformation $A$ from $\mathfrak{H}$ into $\mathfrak{H}$ is termed an operator; its bound is denoted by $\|A\|$. The identity operator will be denoted by $I$. An operator $A$ is invertible if there exists an operator $B$

such that $AB = BA = I$; we then write $B = A^{-1}$. If $\|A\| < 1$, then $I - A$ is invertible and $(I - A)^{-1} = I + A + A^2 + \cdots$ where the last series is convergent even uniformly (in bound). The set of all complex numbers $\lambda$ for which $A - \lambda I$ is not invertible defines the spectrum of $A$, in symbol $\Lambda(A)$; the last is always a closed subset of the circle $|z| \leqq \|A\|$.

The adjoint of $A$ is denoted by $A^*$. The condition $A = A^*$ defines a Hermitean operator. If $A$ is Hermitean, then $\Lambda(A)$ is a subset of the real axis; the proper values of $A$ of infinite multiplicity and the accumulation points of $\Lambda(A)$ define the limit points of the spectrum of $A$. The operator $A$ is positive — in symbol $A \geqq 0$ — if $(Af, f) \geqq 0$ for all $f \in \mathfrak{H}$. Since $\mathfrak{H}$ is assumed to be a complex space, every positive operator $A$ is necessarily Hermitean. An operator $A$ is normal if $A^*A = AA^*$; we reserve the letter $U$ for unitary operators, i. e., such that $U^*U = UU^* = I$, and the letter $E$ or $P$ for projections, that is, such that $E^* = E = E^2$. The projection on a subspace $\mathfrak{M}$ will sometimes be denoted $E_{\mathfrak{M}}$.

An operator $W$ which is isometric on a subspace $\mathfrak{M}$ and equal to 0 on $\mathfrak{H} \ominus \mathfrak{M}$ is defined as *partially isometric*; $\mathfrak{M}$ is the initial set of $W$, while the range $\mathscr{R}$ of $W$ is its final set; we have $W^*W = E_{\mathfrak{M}}$ and $WW^* = E_{\mathscr{R}}$.

For every positive operator $A$ there is one and only one positive operator $B$ such that $A = B^2$; we write $B = A^{\frac{1}{2}}$. In particular, for every operator $A$, the operator $(A^*A)^{\frac{1}{2}}$ is well defined; we find it more convenient to write $[A]$ instead of $(A^*A)^{\frac{1}{2}}$.[1] Clearly, $[A] = [A^*]$ if and only if $A$ is normal. If $A$ is of finite rank (that is, has a finite dimensional range) then both $A^*A$ and $AA^*$ and therefore also $[A]$ and $[A^*]$ are of finite rank.

Theorem. *Let $A$ be an operator. There exists a partially isometric operator $W$ whose initial set is the closure of the range of $[A]$ and the final set is the closure of the range of $A$, satisfying the following relations*

(i) $A = W[A]$,
(ii) $[A] = W^*A$,
(iii) $A^* = W^*[A^*]$,
(iv) $[A^*] = W[A]W^*$.

*The decomposition presented in these formulae is unique in the following sense: If $A = W_1 B_1$ where $B_1 \geqq 0$ and $W_1$ is partially isometric having as its initial set the closure of the range of $B_1$, then $B_1 = [A]$ and $W_1 = W$.*

*In the case $A$ is of finite rank we may assume that $W$ is unitary (not necessarily unique however).*

[1] The symbol $[A]$ introduced above is new. We consider it an improvement over $\mathrm{abs}(A)$ or $|A|$ used sometimes for $(A^*A)^{\frac{1}{2}}$.

*Proof.* First we prove the existence of the above decomposition. It is convenient to replace $[A]$ and $[A^*]$ by $B$ and $C$ respectively. Clearly $B^*B = B^2 = A^*A$ implies that

$$\|Bf\|^2 = (Bf, Bf) = (B^2 f, f) = (A^*Af, f) = \|Af\|^2$$

that is,

$$\|Bf\| = \|Af\| \qquad \text{for all} \quad f \in \mathfrak{H} .$$

In particular, $\|B(f_1 - f_2)\| = 0$ if and only if $\|A(f_1 - f_2)\| = 0$, that is, $Bf_1 = Bf_2$ if and only if $Af_1 = Af_2$. Consequently, the equality

$$Af = W_0(Bf)$$

defines the isometric transformation $W_0$ from $\mathscr{R}_B$ — the range of $B$ onto $\mathscr{R}_A$ — the range of $A$. Clearly $W_0$ may be extended in a unique manner to a partially isometric operator $W$ having the closure of $\mathscr{R}_B$ as its initial set and the closure of $\mathscr{R}_A$ as its final set. Thus,

$$A = WB .$$

(ii): We recall that $W^*W$ is the projection on the initial set of $W$, that is, on $\vee \mathscr{R}_B$, while $WW^*$ is the projection on the final set of $W$, that is, on $\vee \mathscr{R}_A$. Thus,

$$B = W^*W \cdot B = W^* \cdot WB = W^*A .$$

(iv): We have $WBW^* \geqq 0$ since $(WBW^*f, f) = (BW^*f, W^*f) \geqq 0$. Moreover,

$$(WBW^*)^2 = WBW^* \cdot WBW^* = WB \cdot W^*WBW^* =$$
$$= WB \cdot BW^* = AA^* = C^2 .$$

The uniqueness of the square root implies then

$$C = WBW^* .$$

(iii): Multiplying the last equality by $W^*$ on the left, we get

$$W^*C = W^*WBW^* = BW^* = A^* .$$

Uniqueness: Suppose that also $A = W_1 B_1$ where $B_1 \geqq 0$ and $W_1$ is partially isometric having as its initial set the closure of the range of $B_1$. Then, $A^* = B_1 W_1^*$ and

$$B^2 = A^*A = B_1 W_1^* W_1 B_1 = B_1^2 .$$

Thus, $B_1 = B$ and therefore also $W_1 = W$.

Finally, suppose that $A$ is of finite rank. Since $\|Af\| = \|Bf\|$, the ranges of $A$ and $B$ are of the same dimension. Due to this finite equidimensionality, the isometric transformation $W_0$ of all the $Bf$ onto all the $Af$ is one-to-one and possesses a unitary extension (not unique however!). This concludes the proof.

The unique decomposition of an operator stated above will be referred to in the future as its "polar decomposition" (see VON NEUMANN [3]).

Corollary. *The operators $A$ and $[A]$ have the same nullspace. Moreover, $A$ and $[A^*]$ have the same range.*

*Proof.* The first statement is clearly implied by (i) and (ii). Similarly, $A^*$ and $[A^*]$ have the same nullspace. To prove the second statement we avail ourselves of the notation introduced in the proof of the previous theorem. Moreover, let $\mathfrak{N}_A, \mathfrak{N}_{A^*}, \ldots$ denote the nullspace of $A, A^*, \ldots$ respectively.

The partially isometric operator $W$ transforms its initial set

$$\vee \mathscr{R}_B = \mathfrak{H} \ominus \mathfrak{N}_{B^*} = \mathfrak{H} \ominus \mathfrak{N}_B = \mathfrak{H} \ominus \mathfrak{N}_A$$

onto its final set $\vee \mathscr{R}_A = \mathfrak{H} \ominus \mathfrak{N}_{A^*}$; $W$ is 0 on $\mathfrak{N}_A$. On the other hand the operator $[A^*]$ being 0 on $\mathfrak{N}_{A^*}$, assumes its full range on $\mathfrak{H} \ominus \mathfrak{N}_{A^*}$. It follows that $[A^*]$ and $[A^*]W = A$ have the same range.

## 6. Banach spaces

A Banach space is a normed linear space (complex vector space with a norm $\|f\|$) for which the resulting metric space is complete. The set of all bounded linear functionals (complex valued functions) on a normed linear space $\mathscr{B}$ is a Banach space — termed the conjugate space of $\mathscr{B}$ and denoted by $\mathscr{B}^*$ — if the linear operations for functionals are the usual ones and the bound of a functional represents its norm. Let $f_0$ be fixed in $\mathscr{B}$. As $f^*$ varies in $\mathscr{B}^*$, it generates the linear functional $f^*(f_0)$ with a bound equal to $\|f_0\|$, hence an element of $\mathscr{B}^{**}$. Thus a normed linear space admits a "natural" imbedding in its second conjugate space. We express this by writing $\mathscr{B} \subset \mathscr{B}^{**}$. The space $\mathscr{B}$ is termed reflexive (sometimes regular) if in the above imbedding $\mathscr{B}$ coincides with $\mathscr{B}^{**}$. A Banach space is reflexive if and only if its conjugate space is reflexive. A subspace of a reflexive Banach space is also reflexive (compare PETTIS [1] and JAMES [1]).

A mapping $f \to f^*$ on an algebra $\mathscr{A}$ is an involution if it satisfies the following conditions: 1) $(f^*)^* = f$, 2) $(f+g)^* = f^* + g^*$, 3) $(cf)^* = \bar{c} f^*$, 4) $(fg)^* = g^* f^*$. An algebra with an involution is also termed a *-algebra. The existence of an involution guarantees many desirable "symmetries". For instance, to every right ideal corresponds a left ideal via the mapping $f \to f^*$.

A Banach algebra is an algebra for which the underlying linear space is a Banach space and the norm satisfies the extra condition $\|fg\| \leqq \|f\| \, \|g\|$.

Two Banach spaces (or algebras) $\mathscr{B}_1$ and $\mathscr{B}_2$ are termed equivalent if there exists a norm-preserving isomorphism from $\mathscr{B}_1$ onto $\mathscr{B}_2$.

## 7. Remarks on the notation

The notation involved in this exposition is quite clear from the preliminaries and the context. Just the same it seems desirable to list those few symbols which appear frequently.

We denote by $\mathfrak{A}$ the algebra of all operators on $\mathfrak{H}$ while $\mathfrak{R}$ stands for its subalgebra of all operators of finite rank. Frequently, $\mathfrak{R}_n$ will stand for the algebra of all operators on a (finite) $n$-dimensional space $\mathfrak{H}_n$.

Considering in $\mathfrak{A}$ the bound of an operator as its norm, one obtains the Banach algebra $\mathfrak{B}$. We denote by $\mathfrak{C}$ the Banach subalgebra of $\mathfrak{B}$ consisting of all completely continuous operators.

We write $f_n \rightharpoonup f$ to indicate that the sequence of vectors $\{f_n\}$ converges weakly to $f$, that is, $(f_n, g) \to (f, g)$ for every $g \in \mathfrak{H}$; as usual $f_n \to f$ will symbolize (strong) convergence, that is, $\|f_n - f\| \to 0$.

The real and imaginary part of a complex number $\lambda$ is denoted by $\mathscr{R}\lambda$ and $\mathscr{I}\lambda$ respectively. Finally, the symbol $\{x : \pi(x)\}$ represents the set of all $x$'s for which proposition $\pi(x)$ is true.

The book is divided into five chapters. Lemma III, 2 stands for Lemma 2 in chapter III. Throughout chapter III, however, this lemma is briefly referred to as Lemma 2.

# I. The class of operators of the form $\Sigma_j \lambda_j \varphi_j \otimes \overline{\psi}_j$

## 1. The operator $\Sigma_j \lambda_j \varphi_j \otimes \overline{\psi}_j$

Definition 1. Let $\varphi$ and $\psi$ be two given vectors in $\mathfrak{H}$. The symbol

$$\varphi \otimes \overline{\psi}$$

represents then a transformation on $\mathfrak{H}$, whose defining equation is given by

$$(\varphi \otimes \overline{\psi}) f = (f, \psi) \varphi \qquad \text{for } f \in \mathfrak{H}\,.$$

Lemma 1. *The transformation $\varphi \otimes \overline{\psi}$ defined above, is an operator; its bound $\|\varphi \otimes \overline{\psi}\| = \|\varphi\|\,\|\psi\|$. The range of $\varphi \otimes \overline{\psi}$ is of dimension* 1 *or* 0.

Lemma 2. (i) $(\varphi \otimes \overline{\psi})^* = \psi \otimes \overline{\varphi}$.

(ii) $(\lambda\, \varphi) \otimes \overline{\psi} = \lambda(\varphi \otimes \overline{\psi})$.

(ii') $\varphi \otimes \overline{(\lambda\, \psi)} = \overline{\lambda}(\varphi \otimes \overline{\psi})$.

(iii) $(\varphi_1 + \varphi_2) \otimes \overline{\psi} = \varphi_1 \otimes \overline{\psi} + \varphi_2 \otimes \overline{\psi}$.

(iii') $\varphi \otimes \overline{(\psi_1 + \psi_2)} = \varphi \otimes \overline{\psi}_1 + \varphi \otimes \overline{\psi}_2$.

(iv) $(\varphi_1 \otimes \overline{\psi}_1)(\varphi_2 \otimes \overline{\psi}_2) = (\varphi_2, \psi_1)\, \varphi_1 \otimes \overline{\psi}_2$.

(v) $A(\varphi \otimes \overline{\psi}) = (A\,\varphi) \otimes \overline{\psi}$.

(vi) $(\varphi \otimes \overline{\psi})\, A = \varphi \otimes \overline{(A^* \psi)}$.

*Proof.* The above relationships are simple consequences of our definition of $\varphi \otimes \overline{\psi}$.

The meaning of the symbol $\Sigma_{i=1}^{n} \lambda_i \varphi_i \otimes \overline{\psi}_i$ is then clear; it represents an operator of rank at most $n$ (i. e., whose range is at most $n$-dimensional). In general, analogous infinite sums do not make sense. However, the following result in this direction is sufficient for our purposes.

Theorem 1. *Let $\{\varphi_j\}$ and $\{\psi_j\}$ be two orthonormal families of vectors and $\{\lambda_j\}$ a family of complex numbers, indexed by the same set $\{j\}$. The family*

$$\{\lambda_j(f, \psi_j)\varphi_j\} \tag{1}$$

*is summable for every $f \in \mathfrak{H}$ if and only if $\{\lambda_j\}$ is bounded. Whenever $\{\lambda_j\}$ is bounded, the sum $\Sigma_j \lambda_j(f, \psi_j)\varphi_j$ defines an operator which is denoted symbolically by $\Sigma_j \lambda_j \varphi_j \otimes \overline{\psi}_j$. The bound of the last operator is given by*

$$\sup_j |\lambda_j| . \tag{2}$$

*Proof.* Assume first that $\{\lambda_j\}$ is bounded. The summability of the family (1) is obviously equivalent to the summability of the family

$$\{|\lambda_j|^2 \, |(f, \psi_j)|^2\} \tag{3}$$

of non-negative numbers. It is a consequence of Bessel's inequality (for finite orthonormal families of vectors) and of course of the boundedness of $\{\lambda_j\}$ that (3) is always summable and its sum is

$$\leqq \sup_j |\lambda_j|^2 \, \|f\|^2 .$$

For $f \in \mathfrak{H}$, we put then

$$Af = \Sigma_j \lambda_j(f, \psi_j)\varphi_j .$$

Clearly, $A$ is linear. Moreover,

$$\|Af\|^2 = \Sigma_j |\lambda_j|^2 \, |(f, \psi_j)|^2 \leqq \sup_j |\lambda_j|^2 \, \|f\|^2$$

implies that $A$ is an operator with a bound $\leqq \sup_j |\lambda_j|$. On the other hand $\|A\psi_j\| = |\lambda_j|$. Therefore,

$$\|A\| = \|\Sigma_j \lambda_j \varphi_j \otimes \overline{\psi}_j\| = \sup_j |\lambda_j| .$$

Conversely. Suppose that (1) or equivalently (3) is summable for all $f$. If $\{\lambda_j\}$ were not bounded, we could extract a sequence $\{\lambda_{j_n}\}$ such that $|\lambda_{j_n}| \geqq n$ for all $n$. The summability of (3) obviously implies the summability of the sequence $\{|\lambda_{j_n}|^2 \, |(f, \psi_{j_n})|^2\}$ and therefore also of $\{\lambda_{j_n}(f, \psi_{j_n}) \, \varphi_{j_n}\}$ for every $f \in \mathfrak{H}$. Since the summability of a sequence of orthogonal vectors is equivalent to convergence of the series formed by these vectors, we conclude that the sequence of operators $\{T_k\} = \{\Sigma_{n=1}^{k} \lambda_{j_n} \varphi_{j_n} \otimes \overline{\psi}_{j_n}\}$ converges for every $f$. It follows that the corresponding sequence of norms $\{\|T_k\|\}$ is bounded (see BANACH [1, p. 80]). On the other hand, by what was already proven

$$\|T_k\| = \max_{1 \leq n \leq k} |\lambda_{j_n}| \, .$$

We arrive thus at the contradictory statement

$$\|T_k\| \geqq |\lambda_{j_k}| \geqq k \qquad \text{for all } k \, .$$

Corollary. *The operator* $\Sigma_j \lambda_j \varphi_j \otimes \overline{\psi}_j$ *is* 0 *if and only if all the* $\lambda_j$ *are* 0. *Consequently,*

$$\Sigma_j \lambda_j \varphi_j \otimes \overline{\psi}_j = \Sigma_j \mu_j \varphi_j \otimes \overline{\psi}_j \textit{ if and only if } \lambda_j = \mu_j \textit{ for all } j \, .$$

We make the following convention: *Unless the contrary is explicitly stated, it is assumed that in every expression* $\Sigma_j \lambda_j \varphi_j \otimes \overline{\psi}_j$ *considered in the sequel,* $\{\varphi_j\}$ *and* $\{\psi_j\}$ *are two orthonormal families of vectors and* $\{\lambda_j\}$ *is bounded, that is, our expression represents an operator. We shall also assume that all the* $\lambda_j$ *are* $\neq 0$.

Let $A = \Sigma_j \lambda_j \varphi_j \otimes \overline{\psi}_j$. Clearly $Af = 0$ if and only if $(f, \psi_j) = 0$ for all $j$. Thus, $A$ assumes for different vectors different values if and only if $Af = 0$ implies $f = 0$, that is, if $(f, \psi_j) = 0$ for all $j$ implies $f = 0$. The last will happen if and only if $\{\psi_j\}$ is complete. We also notice that the range of $A$ is dense in $\mathfrak{H}$ if and only if $\{\varphi_j\}$ is complete. In particular, when $A$ has an inverse, then necessarily both $\{\varphi_j\}$ and $\{\psi_j\}$ are complete orthonormal families.

We say a few words about the class of operators distinguished by the previous theorem. Consider the operator $A = \Sigma_j \lambda_j \psi_j \otimes \overline{\psi}_j$. It is clear that each $\psi_j$ is a proper vector of $A$ corresponding to the proper value $\lambda_j$. If the $\psi_j$ do not form a complete family, we supplement it to a complete orthonormal family by adding an orthonormal family $\{\omega_i\}$. It is also clear that $\omega_i$ is then a proper vector of $A$ corresponding to the proper value 0. Thus, the operators of the form $A = \Sigma_j \lambda_j \psi_j \otimes \overline{\psi}_j$ are precisely those which have a complete orthonormal family of proper vectors. Next, choose a partially isometric operator $W$ having as its initial set the subspace determined by the $\psi_j$. Denoting $W\psi_j$ by $\varphi_j$, we have $WA = \Sigma_j \lambda_j \varphi_j \otimes \overline{\psi}_j$, and thus *our theorem distinguishes the partially isometric transforms of the operators having a complete orthonormal family of proper vectors*. The "structure of the class" of all operators of the form $WA$ is not simple and will be discussed in another publication.

The proof of the following two theorems is immediate:

Theorem 2. Let $A = \Sigma_j \lambda_j \varphi_j \otimes \overline{\psi}_j$. *Then,*

$$\begin{aligned} A^* &= \Sigma_j \overline{\lambda}_j \psi_j \otimes \overline{\varphi}_j, \\ A^*A &= \Sigma_j |\lambda_j|^2 \psi_j \otimes \overline{\psi}_j, \\ [A] &= \Sigma_j |\lambda_j| \psi_j \otimes \overline{\psi}_j. \end{aligned}$$

*The operator $\Sigma_j \lambda_j \varphi_j \otimes \overline{\varphi}_j$ is normal; it is Hermitean if and only if all the $\lambda_j$ are real.*

Theorem 3. *An operator is*
i) *a projection, if and only if it is of the form*

$$\Sigma_j \varphi_j \otimes \overline{\varphi}_j;$$

*the range of the projection being the subspace determined by $\{\varphi_j\}$.*
ii) *unitary, if and only if it is of the form*

$$\Sigma_j \varphi_j \otimes \overline{\psi}_j$$

*where both $\{\varphi_j\}$ and $\{\psi_j\}$ are complete orthonormal families.*
iii) *partially isometric, if and only if it is of the form*

$$W = \Sigma_j \varphi_j \otimes \overline{\psi}_j$$

*where both $\{\varphi_j\}$ and $\{\psi_j\}$ are orthonormal families (not necessarily complete); its initial and final set are the subspaces determined by the $\psi_j$ and by the $\varphi_j$ respectively. We also have $W^*W = \Sigma_j \psi_j \otimes \overline{\psi}_j$ — the projection on the initial set and $WW^* = \Sigma_j \varphi_j \otimes \overline{\varphi}_j$ — the projection on the final set.*
iv) *isometric, if and only if it is of the form*

$$V = \Sigma_j \varphi_j \otimes \overline{\psi}_j$$

*where $\{\psi_j\}$ and $\{\varphi_j\}$ are orthonormal families and $\{\psi_j\}$ is complete.* It is then clear what is $V^*V$ and $VV^*$.

It is easy to construct examples of Hermitean, even positive operators — of course, on an infinite-dimensional space — which do not have a single proper value. For example, consider $L^2$ and the operator which carries a function $f(x)$ into $xf(x)$. It is also true that such an operator $A \geqq 0$ must not admit a representation $\Sigma_j \lambda_j \varphi_j \otimes \overline{\psi}_j$. Otherwise, we would have $A = [A] = \Sigma_j |\lambda_j| \psi_j \otimes \overline{\psi}_j$; it would then follow that each $|\lambda_j|$ is a proper value of $A$.

## 2. The spectrum of the operator $\Sigma_j \lambda_j \varphi_j \otimes \overline{\varphi}_j$

Theorem 4. *The operator $A = \Sigma_j \lambda_j \varphi_j \otimes \overline{\psi}_j$ has an inverse if and only if*

(i) *$\{\varphi_j\}$ is a complete orthonormal family,*
(ii) *$\{\psi_j\}$ is a complete orthonormal family,*
(iii) *$\{\lambda_j\}$ is not only bounded, but also bounded away from* 0.

*Whenever $A$ has an inverse $A^{-1}$, then*

$$A^{-1} = \Sigma_j \frac{1}{\lambda_j} \psi_j \otimes \overline{\varphi}_j .$$

*Proof.* Assume that the operator $A^{-1}$ exists. The necessity of (i) and (ii) is stated in the remark following Theorem 1. To prove (iii), observe that for every $f \in \mathfrak{H}$,

$$\|f\| = \|A^{-1}Af\| \leqq \|A^{-1}\| \, \|Af\| .$$

Consequently,

$$|\lambda_j| = \|A\psi_j\| \geqq \frac{\|\psi_j\|}{\|A^{-1}\|} = \frac{1}{\|A^{-1}\|} .$$

Conversely. Suppose that (i), (ii), and (iii) hold. Then $\left\{\frac{1}{\lambda_j}\right\}$ is bounded. By Theorem 1, $\Sigma_j \frac{1}{\lambda_j} \psi_j \otimes \overline{\varphi}_j$ defines an operator. Of course,

$$(\Sigma_j \lambda_j \varphi_j \otimes \overline{\psi}_j)\left(\Sigma_j \frac{1}{\lambda_j} \psi_j \otimes \overline{\varphi}_j\right) = \Sigma_j \varphi_j \otimes \overline{\varphi}_j = I$$

and

$$\left(\Sigma_j \frac{1}{\lambda_j} \psi_j \otimes \overline{\varphi}_j\right)(\Sigma_j \lambda_j \varphi_j \otimes \overline{\psi}_j) = \Sigma_j \psi_j \otimes \overline{\psi}_j = I .$$

The representation of an operator in the form of a sum (whenever this is feasible) is very convenient. Suppose that $A = \Sigma_j \lambda_j \varphi_j \otimes \overline{\varphi}_j$. We recall that the spectrum $\Lambda(A)$ of $A$ is the set of all complex numbers $\lambda$ for which $A - \lambda I$ is not invertible. Theorem 4, not only characterizes the spectrum of $A$ but also permits us to derive a *representation for the inverse of $A - \lambda I$ whenever such exists.* We consider two cases:

i) $\{\varphi_j\}$ is complete. Then the spectrum $\Lambda(A)$ of $A$ is the closed subset of the complex plane determined by the $\lambda_j$. For a number $\lambda \notin \Lambda(A)$ we then have,

$$(A - \lambda I)^{-1} = \Sigma_j \frac{1}{\lambda_j - \lambda} \varphi_j \otimes \overline{\varphi}_j .$$

ii) $\{\varphi_j\}$ is not complete. We extend it to a complete orthonormal family by adding $\{\omega_i\}$. Then,

$$A - \lambda I = \Sigma_j (\lambda_j - \lambda) \varphi_j \otimes \overline{\varphi}_j + \Sigma_i (-\lambda) \omega_i \otimes \overline{\omega}_i .$$

The spectrum $\Lambda(A)$ thus consists of the closed subset of the complex plane determined by the $\lambda_j$ plus the number 0. If $\lambda \notin \Lambda(A)$, then by Theorem 4,

$$(A - \lambda I)^{-1} = \Sigma_j \frac{1}{\lambda_j - \lambda} \varphi_j \otimes \overline{\varphi}_j - \frac{1}{\lambda} \Sigma_i \omega_i \otimes \overline{\omega}_i$$

and therefore

$$(A - \lambda I)^{-1} = -\frac{1}{\lambda} I + \frac{1}{\lambda} \cdot \Sigma_j \frac{\lambda_j}{\lambda_j - \lambda} \varphi_j \otimes \overline{\varphi}_j .$$

We remark incidentally, that given a closed and bounded subset $\Lambda$ of the complex plane then a normal operator $A$ (on a given infinite-dimensional space) having $\Lambda$ as its spectrum may be constructed as follows: Let $\lambda_1, \lambda_2, \lambda_3, \ldots$ be a dense sequence in $\Lambda$. Choose an orthonormal sequence $\varphi_1, \varphi_2, \varphi_3, \ldots$ and if necessary extend it to a basis in the space by adding $\{\omega_i\}$. The normal operator

$$A = \Sigma_j \lambda_j \varphi_j \otimes \overline{\varphi}_j + \Sigma_i \lambda_1 \omega_i \otimes \overline{\omega}_i$$

will have $\Lambda$ as its spectrum. In the case $\Lambda$ is on the real axis, $A$ is Hermitean. This is an immediate consequence of Theorem 4.

Corollary. *Whenever $A$ can be represented as a finite or denumerably infinite sum $\Sigma_j \lambda_j \varphi_j \otimes \overline{\varphi}_j$ with real $\lambda_j$ and $\lambda_j \to 0$, then $\Lambda(A)$ consists precisely of all the $\lambda_j$ and 0, (all $\lambda_j$, if the sum has a finite number of terms equal to the dimension of the space). We shall prove later that this is the case if and only if $A$ is completely continuous and Hermitean.*

## 3. Completely continuous operators

This section is devoted to the definition, analysis and a few characterizations of completely continuous operators.

In the literature, one finds several (not equivalent) definitions of a compact set. It is therefore in order to state the sense in which it will be used below:

A subset $\mathfrak{S}$ of a not necessarily complete linear space with an inner product is termed *compact* if every infinite sequence of vectors in $\mathfrak{S}$ contains a subsequence which converges (in the strong sense) to some vector in the space.

It is clear that every compact set is bounded and that the converse is not necessarily true (for example, an infinite orthonormal sequence of vectors). It is also clear that every operator transforms bounded sets into bounded sets, compact sets into compact sets, weakly convergent sequences into such and strongly convergent sequences into such. An operator which does more than that and satisfies any of the two equivalent definitions which follow is termed *completely continuous* (by some writers also "compact"; in French, totalement or complètement continue; in German, vollstetig).

Definition 2. An operator $A$ is completely continuous if it transforms bounded sets into compact sets.

Definition 3. An operator $A$ is completely continuous if it transforms every weakly convergent sequence of vectors into a strongly convergent sequence; symbolically, $f_n \rightharpoonup f$ implies $Af_n \to Af$.

The concept of a completely continuous operator is basically due to Hilbert [1] who is being credited with the second of the above definitions (not quite in the form expressed here). The first definition is due to F. Riesz [1]. The proof of equivalence of the above two definitions is easy and may be found in Riesz and Nagy [1]. It is also clear that every operator of finite rank is necessarily completely continuous.

We denote by $\mathfrak{A}$ the algebra of all operators on $\mathfrak{H}$ and by $\mathfrak{C}$ its subset of all completely continuous operators.

Lemma 3. (i) *$A \in \mathfrak{C}$ implies $\lambda A \in \mathfrak{C}$ ($\lambda$ is complex),*
(ii) *$A \in \mathfrak{C}$ and $B \in \mathfrak{C}$ implies $(A + B) \in \mathfrak{C}$,*
(iii) *$A \in \mathfrak{C}$ and $X \in \mathfrak{A}$ implies $(AX) \in \mathfrak{C}$ and $(XA) \in \mathfrak{C}$.*

*Proof.* This is an immediate consequence of the definition of a completely continuous operator.

Lemma 4. *The following statements are equivalent:*

(i) *$A$ is completely continuous,*
(ii) *$A^*$ is completely continuous,*
(iii) *$[A]$ is completely continuous,*
(iv) *$[A^*]$ is completely continuous.*

*Proof.* This follows at once from the polar decomposition of operators.

Lemma 5. *Let $A_n$ be completely continuous for $n = 1, 2, \ldots$ and $\|A - A_n\| \to 0$. Then $A$ is also completely continuous.*

*Proof.* Let $f_n \rightharpoonup f$. Then for some $c > 0$, $\|f_n\| \leqq c$ for all $n$. The complete continuity of $A_k$ implies: $A_k f_n \to A_k f$. Moreover,

$$\|Af_m - Af_n\| \leqq \|Af_m - A_k f_m\| + \|A_k f_m - A_k f_n\| + \|A_k f_n - Af_n\| \leqq$$
$$\leqq \|A - A_k\| (\|f_m\| + \|f_n\|) + \|A_k f_m - A_k f_n\| .$$

Let $\varepsilon > 0$ be given. Choose a fixed $k$ so that $\|A - A_k\| < \frac{\varepsilon}{4c}$ and then an $N$ so large that $\|A_k f_m - A_k f_m\| < \frac{\varepsilon}{2}$ when both $m$ and $n$ are $\geqq N$. We then have $\|Af_m - Af_n\| < \varepsilon$. Thus, $Af_n \to Af$.

Corollary. *Let $\{\varphi_i\}$ and $\{\psi_i\}$ be two orthonormal sequences of vectors and $\{\lambda_i\}$ a convergent to 0 sequence of complex numbers. Then the operator $\Sigma_i \lambda_i \varphi_i \otimes \overline{\psi}_i$ is completely continuous.*

*Proof.* We have for $n = 1, 2, \ldots$

$$\|\Sigma_i \lambda_i \varphi_i \otimes \overline{\psi}_i - \Sigma_{i=1}^n \lambda_i \varphi_i \otimes \overline{\psi}_i\| = \|\Sigma_{i>n} \lambda_i \varphi_i \otimes \overline{\psi}_i\| = \sup_{i>n} |\lambda_i| .$$

An application of the previous lemma concludes the proof.

Theorem 5. *Consider the set $\mathfrak{C}$ of all completely continuous operators. With the usual definition of sum, product, and scalar multiple for operators, $\mathfrak{C}$ is a self-adjoint two-sided ideal in the algebra $\mathfrak{A}$. The algebra $\mathfrak{C}$ (as well as any algebra of operators) will be normed if the bound of an operator stands for its norm. Moreover, the resulting normed algebra is complete, that is, $\mathfrak{C}$ is a Banach algebra.*

*Proof.* The proof is an immediate consequence of the preceeding three lemmas.

Remark. It is a consequence of the polar decomposition for operators, that any (two-sided) ideal in $\mathfrak{A}$ is necessarily self-adjoint. It will also follow from a later discussion that the bound of an operator is essentially the only norm for which the linear space of *all* completely continuous operators (on an infinite-dimensional Hilbert space) is a Banach algebra.

## 4. The spectral representation of a completely continuous Hermitean operator

The fundamental theorem of algebra implies that the characteristic equation of a complex matrix possesses at least one, in general, complex root. It follows that an operator on a finite-dimensional (complex) space has at least one proper value. Although in general, we can add no more to the above statement, we already have said a lot. The situation is entirely different on an infinite-dimensional space. There, even a completely continuous operator may not admit a proper value. To construct such an example, choose a space with a complete orthonormal sequence of vectors $\{\varphi_i\}$, and a bounded sequence of non-zero numbers $\{\lambda_i\}$. The operator

$$A = \Sigma_i \lambda_i \varphi_{i+1} \otimes \overline{\varphi}_i$$

has no proper value, i. e., the equality

$$Af = \mu f$$

for some complex $\mu$, admits $f = 0$ as the only solution. For $f = \Sigma_i a_i \varphi_i$, our equality amounts to

$$\Sigma_i \lambda_i a_i \varphi_{i+1} = \mu \, \Sigma_i a_i \varphi_i \, .$$

We distinguish two possibilities:

1) $\mu = 0$. Then the left side is also 0. Hence $\lambda_i a_i = 0$ and also $a_i = 0$ (since $\lambda_i \neq 0$).

2) $\mu \neq 0$. Comparing the corresponding coefficients on both sides, we get $\mu a_1 = 0$, $\lambda_1 a_1 = \mu a_2$, $\lambda_2 a_2 = \mu a_3$, .... Hence $a_1 = 0$ and therefore $a_2 = 0$ which again implies $a_3 = 0$, .... We remark finally that whenever also $\lambda_i \to 0$, then the above operator is completely continuous.

In this connection, however, we mention that every completely continuous operator $A$ has a proper invariant subspace. We mean hereby that for some subspace $\mathfrak{M}$, $0 \neq \mathfrak{M} \neq \mathfrak{H}$, we have $A(\mathfrak{M}) \subset \mathfrak{M}$. This non-trivial result is essentially due to J. von Neumann, whose proof has never been published. That the above theorem is valid even for completely continuous operators on an arbitrary Banach space, was shown later by Aronszajn and Smith [1]. We may also add that so far it is not known whether an arbitrary operator even on a Hilbert space has a proper invariant subspace.

The story is different when an operator is also assumed to be Hermitean. Then we can state much more. It is well known that a Hermitean operator $A$ on a finite-dimensional unitary space admits a basis in that space, made up of proper vectors of $A$. This again is a purely algebraic result. Its infinite-dimensional extension — no more purely algebraic — is also well known. It asserts that every Hermitean completely continuous operator on a Hilbert space admits a basis in the space whose elements are proper vectors of $A$. The last assertion and its implications will be of fundamental importance in our discussion: we therefore reconstruct its proof. Since the case $A = 0$ is settled in a trivial manner we shall assume that $A \neq 0$. We recall also that for a Hermitean $A$

$$\|A\| = \sup_{\|f\|=1} |(Af, f)| .$$

Lemma 6. *Let $A$ be Hermitean. A vector $\varphi$ with $\|\varphi\| = 1$ maximizes the right side, that is,*

$$|(A\varphi, \varphi)| = \|A\|$$

*if and only if*

$$A\varphi = \lambda\varphi ,$$

*where $\lambda$ is equal to either $+\|A\|$ or $-\|A\|$.*

*Proof.* Assume that for $\varphi$ with $\|\varphi\| = 1$ we have $|(A\varphi, \varphi)| = \|A\|$. Choosing $\lambda$ so that $\lambda(A\varphi, \varphi) = \|A\|^2$ we have,

$$0 \leqq \|A\varphi - \lambda\varphi\|^2 = \|A\varphi\|^2 - 2\lambda(A\varphi, \varphi) + \lambda^2 \leqq \|A\|^2 - 2\|A\|^2 + \|A\|^2 = 0 .$$

The converse is obvious.

Lemma 7. *Let $A \neq 0$ be a completely continuous Hermitean operator. Then there exists a vector $\varphi_1$ such that $\|\varphi_1\| = 1$ and $A\varphi_1 = \lambda_1\varphi_1$, where $\lambda_1$ is equal to either $+\|A\|$ or $-\|A\|$. Thus, every completely continuous Hermitean operator $A \neq 0$ possesses at least one (necessarily real) proper value $\neq 0$.*

*Proof.* We exhibit a vector $\varphi_1$ maximizing $|(Af, f)|$ if $\|f\| = 1$. Choose a sequence $\{f_n\}$ with $\|f_n\| = 1$ for which

$$|(Af_n, f_n)| \to \|A\| .$$

If necessary, passing to a subsequence, we may suppose also that $(Af_n, f_n)$ converges; its limit $\lambda_1$ is then equal to $+\|A\|$ or $-\|A\|$. Now

$$0 \leqq \|Af_n - \lambda_1 f_n\|^2 = \|Af_n\|^2 - 2\lambda_1(Af_n, f_n) + \lambda_1^2 \leqq \lambda_1^2 - 2\lambda_1(Af_n, f_n) + \lambda_1^2 ,$$

and the extreme right of the last inequality converges to 0. Thus,

$$Af_n - \lambda_1 f_n \to 0 .$$

The complete continuity of $A$ implies that $\{Af_n\}$ contains a convergent subsequence $\{Af_{n_k}\}$. But then $\{f_{n_k}\}$ must converge; denote its limit by $\varphi_1$. Thus,

$$Af_{n_k} - \lambda_1 f_{n_k} \to A\varphi_1 - \lambda_1\varphi_1$$

and therefore

$$A\varphi_1 - \lambda_1\varphi_1 = 0 .$$

Theorem 6. *Every Hermitean completely continuous operator $A$ admits a complete orthonormal family of proper vectors. Its non-zero (necessarily real) proper values are of finite multiplicity and form either a finite or a denumerably infinite sequence $\lambda_1, \lambda_2, \lambda_3, \ldots$;*) in the last case we also have $\lambda_i \to 0$. If, $\varphi_1, \varphi_2, \varphi_3, \ldots$ is a corresponding orthonormal sequence of proper vectors (i. e., $A\varphi_i = \lambda_i\varphi_i$ for $i = 1, 2, \ldots$), then*

$$A = \Sigma_i \lambda_i \varphi_i \otimes \overline{\varphi}_i .$$

*Proof.* Let $\varphi_1$ be a solution of the extremal problem discussed in the previous lemma and $\mathfrak{H}_1$ the subspace orthogonal to $\varphi_1$. The restriction $A_1$ of $A$ to $\mathfrak{H}_1$, has its range in $\mathfrak{H}_1$ and thus may be considered as a completely continuous Hermitean operator on $\mathfrak{H}_1$. If $A_1 \neq 0$, a repetition of the argument employed in the preceding lemma yields a $\varphi_2$ which maximizes $|(Af, f)|$ subject to the conditions $\|f\| = 1$ and $(f, \varphi_1) = 0$. Moreover,

$$A\varphi_2 = \lambda_2\varphi_2$$

where $\lambda_2$ is either $+\|A_1\|$ or $-\|A_1\|$, hence $|\lambda_1| \geqq |\lambda_2|$. In the case $A$ is not identically 0 on the subspace orthogonal to both $\varphi_1$ and $\varphi_2$ we construct again a vector $\varphi_3$ which maximizes $|(Af, f)|$ subject to the conditions $\|f\| = 1$, $(f, \varphi_1) = (f, \varphi_2) = 0$. We also have, $A\varphi_3 = \lambda_3\varphi_3$ with $|\lambda_2| \geqq |\lambda_3|$. Continuing this procedure we obtain a finite or infinite orthonormal sequence $\varphi_1, \varphi_2, \varphi_3, \ldots$ of proper vectors of $A$ corresponding to the non-zero proper values $\lambda_1, \lambda_2, \lambda_3, \ldots$; we have $|\lambda_1| \geqq |\lambda_2| \geqq |\lambda_3| \geqq \cdots$. In the case of an infinite sequence, $\left\{\frac{1}{\lambda_i}\varphi_i\right\}$ must not be bounded since $\left\{A\left(\frac{1}{\lambda_i}\varphi_i\right)\right\} = \{\varphi_i\}$ is obviously not compact. Thus necessarily

$$\lambda_i \to 0 .$$

In the case of a finite sequence $\lambda_1, \ldots, \lambda_n$ we have $A = 0$ on the orthogonal complement spanned by $\varphi_1, \ldots, \varphi_n$. If $\lambda_1, \lambda_2, \ldots$ is an infinite sequence, then $A\varphi = \mu\varphi$ for some $\varphi$ with $\|\varphi\| = 1$ and $(\varphi, \varphi_i) = 0$ for $i = 1, 2, \ldots$ implies $|\mu| = |(A\varphi, \varphi)| \leqq |\lambda_i|$ and thus $\mu = 0$. It follows that the so constructed finite or infinite sequence $\lambda_1, \lambda_2, \ldots$ contains *all the non-zero proper values* of $A$; each proper value appearing in the sequence the number of times equal to its multiplicity. Since $\lambda_i \to 0$, the multiplicity of each proper value is necessarily finite.

Now, let $\{\psi_i\}$ denote *any* corresponding orthonormal sequence of proper vectors, that is, $A\psi_i = \lambda_i\psi_i$ for $i = 1, 2, \ldots$. The subspace $\mathfrak{H}'$ spanned by the $\psi_i$ reduces $A$. Moreover, $A = 0$ on $\mathfrak{H} \ominus \mathfrak{H}'$ in view of Lemma 7. Thus, $A = \Sigma_i \lambda_i \psi_i \otimes \overline{\psi}_i$.

*) A non-zero proper value of multiplicity $n$ is repeated in the sequence precisely $n$ times.

We extend $\{\psi_i\}$ to a complete orthonormal family in our space by adding an orthonormal family $\{\omega_j\}$. Then $A\,\omega_j = 0$ for all $j$, that is, $\omega_j$ is a proper vector of $A$ corresponding to the proper value 0.

The representation $A = \Sigma_i \lambda_i \varphi_i \otimes \bar{\varphi}_i$ derived above for a completely continuous Hermitean operator $A$, will be referred to in the future as its *spectral representation.*

The above spectral representation may be stated in a slightly different form: *Let $\mu_1, \mu_2, \ldots$ denote all the different proper values (among them also* 0 *if the last is also a proper value) of $A$ and $P_j$ stand for the projection on the characteristic subspace corresponding to $\mu_j$. Then*

$$\Sigma_j P_j = I \quad \textit{and} \quad A = \Sigma_j \mu_j P_j .$$

*We also have,*

$$A\,P_j = P_j A = \mu_j P_j$$

*for all $j$. An operator $B$ commutes with $A$ if and only if $B P_j = P_j B$ for all $j$.*

It remains to verify the last statement. Assume thus that $A B = B A$. Since $\mu_j B P_j = B A P_j = A B P_j$, the range of $B P_j$ is contained in the characteristic subspace of $\mu_j$, that is, in the range of $P_j$. Thus, $P_i B P_j = \delta_{ij} B P_j$ where $\delta_{ij} = 0$ if $i \neq j$ and 1 if $i = j$. Hence,

$$P_j B = P_j B(\Sigma_i P_i) = \Sigma_i P_j B P_i = \Sigma_i \delta_{ji} B P_i = B P_j .$$

The conclusion in the other direction is obvious.

Quite often, however, the spectral representation of a completely continuous Hermitean operator $A$ is stated in still a different form: For any real $\lambda$ define

$$E_\lambda = \Sigma_{\mu_j \leq \lambda} P_j .$$

Then $\{E_\lambda\}$ is a resolution of the identity and

$$A = \int_{m-0}^{M} \lambda\, d E_\lambda$$

where $m$ and $M$ stand for the greatest lower and the least upper bound of $A$.

Corollary. *Two completely continuous Hermitean operators $A$ and $B$ commute if and only if there is in our space a basis whose elements are proper vectors of both $A$ and $B$.*

*Proof.* Assume first that the elements in a basis $\{\varphi_j\}$ are proper vectors of both $A$ and $B$. Let

$$A\,\varphi_j = \lambda_j \varphi_j \quad \text{and} \quad B\,\varphi_j = \mu_j \varphi_j .$$

Then,

$$A = \Sigma_j \lambda_j \varphi_j \otimes \overline{\varphi}_j \quad \text{and} \quad B = \Sigma_j \mu_j \varphi_j \otimes \overline{\varphi}_j .$$

Thus,

$$A B = B A = \Sigma_j \lambda_j \mu_j \varphi_j \otimes \overline{\varphi}_j .$$

Conversely. Let $\Sigma_j \zeta_j P_j$ be the spectral form of $A$. Then $AB = BA$ is equivalent to $BP_j = P_j B$ for all $j$. This, of course, means that the range $\mathfrak{M}_j$ of $P_j$ reduces $B$. The restriction of $B$ to $\mathfrak{M}_j$ is again Hermitean and thus there is in $\mathfrak{M}_j$ a basis made up of proper vectors of $B$, which of course are also proper vectors of $A$. This is true for all $j$. The relation

$$\mathfrak{M}_1 \oplus \mathfrak{M}_2 \oplus \cdots = \mathfrak{H}$$

furnishes the desired conclusion.

Remark. *Let $A$ be Hermitean. If for some natural number $p$ the operator $A^p$ is completely continuous, then $A$ is also completely continuous.*

To see this observe first that for any Hermitean $A$ and $g \in \mathfrak{H}$ we have

$$\|A^{\frac{p}{2}} g\|^2 = (A^{\frac{p}{2}} g, A^{\frac{p}{2}} g) = (A^p g, g) \leqq \|A^p g\| \, \|g\|$$

or

$$\|A^{\frac{p+1}{2}} g\|^2 = (A^{\frac{p+1}{2}} g, A^{\frac{p+1}{2}} g) = (A^p g, Ag) \leqq \|A^p g\| \, \|Ag\|$$

depending whether $p$ is even or odd.

Assume now $f_n \rightharpoonup f$. We wish to prove $Af_n \to Af$. Clearly, for some constant $c$ one has $\|f_n\| \leqq c$ for $n = 1, 2, \ldots$ and thus also $\|f\| \leqq c$. Replacing in the above inequalities $g$ by $f_n - f$ one has

$$\|A^{\frac{p}{2}} (f_n - f)\|^2 \leqq \|A^p (f_n - f)\| \, 2c$$

if $p$ is even and

$$\|A^{\frac{p+1}{2}} (f_n - f)\|^2 \leqq \|A^p (f_n - f)\| \, \|A\| \, 2c$$

if $p$ is odd.

The complete continuity of $A^p$ gives $A^p (f_n - f) \to 0$. Hence also $A^{\frac{p}{2}} (f_n - f) \to 0$ or $A^{\frac{p+1}{2}} (f_n - f) \to 0$ by the first or the second of the above inequalities depending whether $p$ is even or odd. Repeating the argument a finite number of times we arrive at $A(f_n - f) \to 0$.

The above remark is also true if $A$ is normal and $A^p$ is completely continuous. Then

$$[A]^{2p} = (A^* A)^p = (A^*)^p A^p$$

is also completely continuous. By what was already shown $[A]$ and therefore also $A$ is completely continuous.

Frequently we shall also use a representation similar to that in Theorem 6 and applicable to *all* completely continuous operators. The last is derived from the spectral representation via the polar decomposition. For this reason we shall refer to it in the future as the *polar representation* of completely continuous operators.

Theorem 7. *The completely continuous operators $A$ are precisely those admitting a "polar representation" $A = \Sigma_i \lambda_i \varphi_i \otimes \overline{\psi}_i$ where both $\{\varphi_i\}$ and $\{\psi_i\}$ are orthonormal sequences and all the $\lambda_i$'s are positive. The sum has* *either a finite or a denumerably infinite number of terms; in the last case, we have also* $\lambda_i \to 0$. *The above representation is unique in the sense that the* $\lambda_i$*'s are necessarily all the positive proper values (each appearing the number of times equal to its multiplicity) of* $[A]$.

*Proof.* Assume that $A$ is completely continuous. Using the polar decomposition $A = W[A]$ and $[A] = W^*A$, we see that this will happen if and only if $[A]$ is completely continuous. The last operator being also positive admits a spectral representation

$$[A] = \Sigma_i \lambda_i \psi_i \otimes \overline{\psi}_i \quad \text{with} \quad \lambda_i > 0 \quad \text{and} \quad \lambda_i \to 0 .$$

Consequently,

$$A = W[A] = \Sigma_i \lambda_i (W\psi_i) \otimes \overline{\psi}_i .$$

Since $W$ is isometric on the subspace determined by the range of $[A]$, the sequence $\{W\psi_i\}$ is also orthonormal. We put $W\psi_i = \varphi_i$ and obtain the desired representation.

Conversely. The corollary to Lemma 5 states that every such representation determines a completely continuous operator.

Uniqueness: Assume that $A = \Sigma_i \mu_i \varphi_i' \otimes \overline{\psi}_i'$ where $\mu_i > 0$ and both $\{\varphi_i'\}$ and $\{\psi_i'\}$ are orthonormal families. Then $[A] = \Sigma_i \mu_i \psi_i' \otimes \overline{\psi}_i'$; the $\mu_i$'s are thus necessarily all the positive proper values of $[A]$.

It may be in order to say a few words about the range of a completely continuous operator $A = \Sigma_i \lambda_i \varphi_i \otimes \overline{\psi}_i$. That range is obviously dense in the separable subspace determined by the sequence $\{\varphi_i\}$. More precisely the range of $A$ is the linear manifold consisting of the sums of the form $\Sigma_i \lambda_i a_i \varphi_i$ where $\{a_i\}$ runs through all sequences of complex numbers subject to the sole restriction $\Sigma_i |a_i|^2 < +\infty$. It is interesting that the above linear manifold must not contain a (closed) subspace of infinite dimension.

Theorem 8. *Let $A$ be completely continuous. Then any subspace in its range is necessarily of finite dimension.*

*Proof.* Denote by $\mathfrak{D}$ and $\mathfrak{R}$ the nullspace and the range of $A$. The restriction of $A$ to $\mathfrak{H} \ominus \mathfrak{D}$ is obviously a completely continuous operator from $\mathfrak{H} \ominus \mathfrak{D}$ into $\mathfrak{H}$ with the same range $\mathfrak{R}$ and which in addition is one-to-one. Let $A^{-1}$ stand for its inverse mapping from $\mathfrak{R}$ onto $\mathfrak{H} \ominus \mathfrak{D}$ (here the symbol $A^{-1}$ does not necessarily stand for an operator). If $\mathfrak{M}$ is a (closed) subspace in $\mathfrak{R}$, then $\mathfrak{N} = A^{-1}\mathfrak{M}$ is also closed (since $A$ is continuous). Since $A$ maps the complete space $\mathfrak{N}$ onto $\mathfrak{M}$ in a one-to-one manner, it follows that $A^{-1}$ is also bounded on $\mathfrak{M}$ (see BANACH [1, p. 41]). A bounded set $\mathfrak{S} \subset \mathfrak{M}$ is transformed by $A^{-1}$ into a bounded set in $\mathfrak{N}$; the last is transformed by the completely continuous operator $A$ back into $\mathfrak{S}$ as a compact set. Thus every bounded set in $\mathfrak{M}$ is necessarily compact. It is well known that the last condition holds if and only if $\mathfrak{M}$ is finite-dimensional.

Theorem 9. *A Hermitean operator $A$ on an infinite-dimensional space is completely continuous if and only if 0 is the only limit point of its spectrum.*

*Proof.* That the spectrum of a Hermitean completely continuous operator $A$ has 0 as its only limit point follows from Theorem 6 and the corollary to Theorem 4.

For the converse we do not choose the simplest route. Instead we use this opportunity to restate that the spectral (integral) representation of a Hermitean operator generalizes the one stated in Theorem 6 (compare NAGY [1] p. 56).

Assume thus that the Hermitean operator

$$A = \int \lambda \, dE_\lambda$$

has 0 as the only limit point of its spectrum. Then any closed interval away from 0 contains at most a finite number of proper values of $A$, and each of these proper values is necessarily of finite multiplicity. All the non-zero proper values of $A$ may thus be arranged in a sequence $\lambda_1, \lambda_2, \ldots$ whose terms are all different and such that $|\lambda_1| \geqq |\lambda_2| \geqq \cdots$. Furthermore, if 0 happens to be also a proper value for $A$, we put $\lambda_0 = 0$.

We find it convenient to introduce the notation

$$E(\lambda) = E_\lambda - E_{\lambda-0} \, .$$

Expressing $\|Af - \mu f\|^2$ and $\|f - E(\mu) f\|^2$ in terms of the above integral, one sees that either both are 0 or both are $\neq 0$. This means

$$Af = \mu f \quad \text{if and only if} \quad f = E(\mu) f \, ;$$

the range of $E(\mu)$ is thus the characteristic subspace corresponding to $\mu$. Of course, we also have

$$E_\lambda = \Sigma_{\lambda_i \leqq \lambda} E(\lambda_i) \, .$$

With these preliminary remarks in mind we are ready for the converse. Assume thus $f_n \rightharpoonup f$. We shall prove $Af_n \to Af$. Clearly,

1) For some constant $c$, we have $\|f_n\| \leqq c$ and $\|f\| \leqq c$.

2) Since the range of $E(\lambda_i)$ is finite dimensional we necessarily have $E(\lambda_i) f_n \to E(\lambda_i) f$.

3) $$\|Af_n - Af\|^2 = \int \lambda^2 \, d\|E_\lambda (f_n - f)\|^2 = \Sigma_i \lambda_i^2 \|E(\lambda_i)(f_n - f)\|^2 = \\ = \Sigma_{i=1}^{r-1} \lambda_i^2 \|E(\lambda_i)(f_n - f)\|^2 + \Sigma_{i \geqq r} \lambda_i^2 \|E(\lambda_i)(f_n - f)\|^2 .$$

Let $\varepsilon > 0$ be given. We examine the sums on the extreme right of the last equality. The second sum is obviously $\leqq \lambda_r^2 \|f_n - f\|^2 \leqq \lambda_r^2 (2c)^2$, hence $< \frac{\varepsilon^2}{2}$ for a sufficiently large $r$. The first sum consists only of $r - 1$ terms; it therefore will be $< \frac{\varepsilon^2}{2}$ for sufficiently large $n$, say all $n \geqq N$. Therefore, for all $n \geqq N$ we have $\|Af_n - Af\| < \varepsilon$.

## 5. Some equalities and inequalities for the proper values of completely continuous Hermitean operators

Consider a Hermitean operator in the form $A = \Sigma_j \lambda_j \varphi_j \otimes \overline{\varphi}_j$; all the $\lambda_j$ are $\neq 0$ and necessarily real. We have,

$$(Af, f) = \Sigma_j \lambda_j |(f, \varphi_j)|^2 .$$

Clearly $A \geqq 0$ if and only if all the $\lambda_j$ are positive. In general, when the $\lambda_j$'s are of different signs, $A$ may be written in the form $A = A_1 - A_2$ where both $A_1$ and $A_2$ are positive; one merely defines $A_1$ as the sum of the terms $\lambda_i \varphi_i \otimes \overline{\varphi}_i$ with $\lambda_i > 0$ and similarly $A_2$ as the sum of the terms $(-\lambda_i)\, \varphi_i \otimes \overline{\varphi}_i$ with $\lambda_i < 0$; either of the sums may be finite or "empty".

Assume now that $A$ is a positive completely continuous operator. We may arrange its positive proper values in a non-increasing sequence $\lambda_1, \lambda_2, \ldots$ with each proper value occurring in the sequence the number of times equal to its multiplicity. Let $\varphi_1, \varphi_2, \ldots$ be a corresponding orthonormal sequence of proper vectors. Then $A = \Sigma_i \lambda_i \varphi_i \otimes \overline{\varphi}_i$. It is also clear from the above equality for $(Af, f)$ that

$$\lambda_n = \max_{||f|| = 1;\, (f, \varphi_i) = 0 \text{ for } i = 1, 2, \ldots, n-1} (Af, f) .$$

The last formula for $\lambda_n$ is not particularly convenient since it expresses $\lambda_n$ via $\varphi_1, \ldots, \varphi_{n-1}$.

The well-known theorem on the *direct characterization of the n-th positive and negative proper value of a Hermitean operator* on a finite-dimensional space was stated first by E. FISCHER [1] and rediscovered later by COURANT [1]. It was then extended by COURANT [2] to a special integral operator with a Hermitean kernel, but actually it carries over in a straightforward manner to any Hermitean completely continuous operator $A$ on a Hilbert space. If in addition $A \geqq 0$, it states:

$$\lambda_n = \min_{\mathfrak{M} \text{ is } (n-1)\text{-dimen.}} \left\{ \max_{||f|| = 1 \text{ and } f \text{ orthogonal to } \mathfrak{M}} (Af, f) \right\}$$

where the last equality is to be interpreted as follows: *Let $\mathfrak{M}$ be any $(n-1)$-dimensional subspace and* $\max(\mathfrak{M})$ *stand for the maximum that $(Af, f)$ assumes as $f$ is orthogonal to $\mathfrak{M}$ and $\|f\| = 1$. Then $\lambda_n$ equals to the minimum of the numbers* $\max(\mathfrak{M})$ *formed for all $(n-1)$-dimensional subspaces $\mathfrak{M}$.*

To prove this, denote by $\mathfrak{M}_n$ the subspace spanned by $\varphi_1, \ldots, \varphi_n$. Whenever $f$ is orthogonal to $\mathfrak{M}_{n-1}$ we have

$$(Af, f) = \Sigma_{i \geqq n} \lambda_i |(f, \varphi_i)|^2 \leqq \lambda_n \|f\|^2 .$$

But $(A\varphi_n, \varphi_n) = \lambda_n$. Thus, $\max(\mathfrak{M}_{n-1}) = \lambda_n$.

On the other hand, given an $(n-1)$-dimensional subspace $\mathfrak{M}$, there is always a vector $\varphi \in \mathfrak{M}_n$ of norm 1 which is orthogonal to $\mathfrak{M}$. For this $\varphi$, $(A\varphi, \varphi) = \Sigma_{i=1}^n \lambda_i |(\varphi, \varphi_i)|^2 \geqq \lambda_n \Sigma_{i=1}^n |(\varphi, \varphi_i)|^2 = \lambda_n$.

Thus, $\max(\mathfrak{M}) \geqq \lambda_n$.

To construct such a vector $\varphi$, one chooses a basis $\omega_1, \ldots, \omega_{n-1}$ in $\mathfrak{M}$ and solves the system of linear equations

$$\Sigma_{i=1}^n x_i(\varphi_i, \omega_j) = 0; \quad j = 1, \ldots, n-1 .$$

If $a_1, \ldots, a_n$ with $\Sigma_{i=1}^n |a_i|^2 = 1$ is a solution of the above system, then $\varphi = a_1\varphi_1 + \cdots + a_n\varphi_n$ satisfies our requirements.

With the aid of the above characterization of proper values in terms of maxima and minima, we prove two lemmas needed in our later discussion.

Lemma 8. *Let $A$ and $X$ be two operators of which $A$ is completely continuous. Let $\lambda_1 \geqq \lambda_2 \geqq \cdots$ and $\mu_1 \geqq \mu_2 \geqq \cdots$ denote the sequences of all positive proper values of $[A]$ and $[XA]$ respectively. If either of the two sequence is finite, we extend it to an infinite sequence by defining the missing terms as 0. Then,*

$$\mu_n \leqq \|X\| \, \lambda_n .$$

*Proof.* The $\lambda_n^2$ are the non-zero proper values of $A^*A$. Hence

$$\lambda_n^2 = \min\{\max(A^*Af, f)\}$$

and since $(A^*Af, f) = \|Af\|^2$

$$\lambda_n = \min\{\max\|Af\|\} .$$

Similarly,

$$\mu_n = \min\{\max\|XAf\|\} .$$

Since $\|XAf\| \leqq \|X\| \, \|Af\|$, the conclusion follows.

Lemma 9. *Let $A$ and $B$ be two completely continuous operators. Let $\lambda_1 \geqq \lambda_2 \geqq \cdots$; $\mu_1 \geqq \mu_2 \geqq \cdots$; $\zeta_1 \geqq \zeta_2 \geqq \cdots$ be the sequences of all the positive proper values of $[A]$, $[B]$ and $[A+B]$ respectively (the number of repetitions of each proper value is equal to its multiplicity). If either of the above sequences is finite, define the missing terms as zero. Then*

$$\zeta_{2i-1}^2 \leqq 2\,(\lambda_i^2 + \mu_i^2)$$

*and therefore also*

$$\zeta_{2i} \leqq \zeta_{2i-1} \leqq 2(\lambda_i + \mu_i); \qquad i = 1, 2, 3, \ldots .$$

*Proof.* Recalling that $[A]^2$ stands for $A^*A$, one readily verifies that

$$[A+B]^2 + [A-B]^2 = 2[A]^2 + 2[B]^2$$

and therefore,

$$[A+B]^2 \leqq 2[A]^2 + 2[B]^2 .$$

Let $N$ be a fixed natural number. Let $\varphi_1, \ldots, \varphi_{N-1}$ denote proper vectors of $[A]$ corresponding to the proper values $\lambda_1, \ldots, \lambda_{N-1}$. Then $\varphi_1, \ldots, \varphi_{N-1}$ are also proper vectors of $[A]^2$ corresponding to the proper values $\lambda_1^2, \ldots, \lambda_{N-1}^2$. As was already pointed out before,

$$\lambda_N^2 = \max_{\|f\|=1;\,(f,\varphi_i)=0 \text{ for } i=1,2,\ldots,N-1} ([A]^2 f, f).$$

Similarly, if $\psi_1, \ldots, \psi_{N-1}$ are proper vectors of $[B]$ corresponding to the proper values $\mu_1, \ldots, \mu_{N-1}$ we have

$$\mu_N^2 = \max_{\|f\|=1;\,(f,\psi_i)=0 \text{ for } i=1,2,\ldots,N-1} ([B]^2 f, f).$$

Since the subspace spanned by $\varphi_1, \ldots, \varphi_{N-1}, \psi_1, \ldots, \psi_{N-1}$ is at most $(2N-2)$-dimensional, the theorem of COURANT referred to before implies

$$\begin{aligned}
\zeta_{2N-1}^2 &\leqq \max_{\|f\|=1;\,(f,\varphi_i)=(f,\psi_i)=0 \text{ for } i=1,2,\ldots,N-1} ([A+B]^2 f, f) \leqq \\
&\leqq 2 \max_{\|f\|=1;\,(f,\varphi_i)=(f,\psi_i)=0 \text{ for } i=1,2,\ldots,N-1} ([A]^2 f, f) + \\
&+ 2 \max_{\|f\|=1;\,(f,\varphi_i)=(f,\psi_i)=0 \text{ for } i=1,2,\ldots,N-1} ([B]^2 f, f) \leqq \\
&\leqq 2 \max_{\|f\|=1;\,(f,\varphi_i)=0 \text{ for } i=1,2,\ldots,N-1} ([A]^2 f, f) + \\
&+ 2 \max_{\|f\|=1;\,(f,\psi_i)=0 \text{ for } i=1,2,\ldots,N-1} ([B]^2 f, f) = 2\lambda_N^2 + 2\mu_N^2.
\end{aligned}$$

## 6. Some ideals of operators

Let $\mathfrak{H}$ be a Hilbert space. The symbols $\mathfrak{A}$ and $\mathfrak{C}$ were already reserved for the algebra of all operators and its subalgebra of all completely continuous operators on $\mathfrak{H}$. We also reserve the symbol $\mathfrak{R}$ for the set of all operators on $\mathfrak{H}$ of finite rank.

Throughout this exposition the term "ideal" will always mean a "two-sided" ideal. Only in section 8 shall we consider left ideals. An ideal in a given algebra will be termed non-trivial or proper if it is not the zero-ideal and it does not embrace the whole algebra. It is of great interest indeed that both $\mathfrak{A}$ and $\mathfrak{C}$ possess non-trivial ideals if and only if $\mathfrak{H}$ is infinite-dimensional.

Lemma 10. *An ideal $\mathfrak{T} \neq 0$ in the algebra of linear transformations on a finite-dimensional space, necessarily coincides with the whole algebra.*

Proof. Let $f_1, \ldots, f_n$ be a basis in our space. Given a vector $f$, let $T_{f,f_i}$ stand for the unique linear transformation assuming the value $f$ for $f_i$ and the value 0 for $f_j$ with $j \neq i$.

Let $0 \neq A \in \mathfrak{T}$. Then for some $f \neq 0$ we have $Af = g \neq 0$. Let $h$ be a given vector and $B$ a linear transformation such that $Bg = h$. Clearly, $BA\,T_{f,f_i} = T_{h,f_i}$. Thus $T_{h,f_i} \in \mathfrak{T}$. Since every linear transformation $T$ may be written in the form

$$T = T_{h_1,f_1} + T_{h_2,f_2} + \cdots + T_{h_n,f_n}$$

our conclusion follows.

The situation is strikingly different when $\mathfrak{H}$ is infinite-dimensional. In this case $\mathfrak{R}$ is always a non-trivial ideal in $\mathfrak{A}$. Moreover $\mathfrak{R}$ is also "absolutely minimal" in the sense expressed below.

Lemma 11. *Let $\mathfrak{H}$ be infinite-dimensional. Then any non-zero ideal $\mathfrak{T}$ in $\mathfrak{A}$ includes $\mathfrak{R}$.*

*Proof.* It is clear that $\mathfrak{R}$ is an ideal in $\mathfrak{A}$. Suppose now that $A \neq 0$ belongs to an ideal $\mathfrak{T}$. Then for some $f \neq 0$ we have $Af = g \neq 0$. Let $\varphi$ and $\psi$ be any two vectors in $\mathfrak{H}$ and $B$ an operator for which $Bg = \varphi$. Clearly,

$$BA(f \otimes \overline{\psi}) = (BAf) \otimes \overline{\psi} = \varphi \otimes \overline{\psi}\,.$$

Thus, $\mathfrak{T}$ contains $\varphi \otimes \overline{\psi}$ and consequently it includes all the operators of finite rank.

The non-existence or existence of non-trivial ideals in $\mathfrak{A}$ is one of the properties which differentiates sharply between a finite and an infinite-dimensional space. Moreover, in the last case we can always construct an "absolutely maximal" non-trivial ideal in $\mathfrak{A}$ i. e., one which contains any other proper ideal in $\mathfrak{A}$. The construction follows:

Lemma 12. *Let $A$ be an operator on $\mathfrak{H}$ containing in its range a (closed) subspace $\mathfrak{M}$ of the same dimension as $\mathfrak{H}$. Then there exist two partially isometric operators $X$ and $Y$ such that $YAX$ has an inverse.*

*Proof.* Let $\mathfrak{N}$ be the null-space of $A$. The restriction $A_1$ of $A$ to $\mathfrak{H} \ominus \mathfrak{N}$ is an operator from $\mathfrak{H} \ominus \mathfrak{N}$ into $\mathfrak{H}$, one-to-one, and with the same range. If $A_1^{-1}$ stands for the inverse mapping of $A_1$, then $A_1^{-1}\mathfrak{M} = \mathfrak{S}$ is a (closed) subspace of $\mathfrak{H} \ominus \mathfrak{N}$.

Choose any two isometric operators $X$ and $Y^*$ from $\mathfrak{H}$ onto $\mathfrak{S}$ and from $\mathfrak{H}$ onto $\mathfrak{M}$ respectively. This, of course, can be done since by assumption $\mathfrak{M}$ and consequently also $\mathfrak{S}$ has the same dimension as $\mathfrak{H}$. The operator $Y = Y^{**}$ is then partially isometric, has $\mathfrak{M}$ as its initial set and $\mathfrak{H}$ as its final set. Consequently, the operator $YAX$ has all of $\mathfrak{H}$ as its range. It is also one-to-one for the following reason: For $f \in \mathfrak{H}$, $Xf \in \mathfrak{S}$ and $AXf \in \mathfrak{M}$. Since $Y$ is isometric on $\mathfrak{M}$, $YAXf = 0$ implies $AXf = 0$ and thus $Xf \in \mathfrak{N}$. Since also $Xf \in \mathfrak{S} \subset \mathfrak{H} \ominus \mathfrak{N}$, we have $Xf = 0$ and consequently also $f = 0$ ($X$ was chosen isometric on $\mathfrak{H}$). The interior mapping principle (see Banach [1, p. 41]) then implies that the operator $YAX$ being one-to-one and having all of $\mathfrak{H}$ as its range, admits a continuous inverse.

Corollary. *No proper ideal $\mathfrak{T}$ in the algebra $\mathfrak{A}$ includes an operator whose range contains a subspace of the same dimension as $\mathfrak{H}$.*

*Proof.* Otherwise, by the previous lemma $\mathfrak{T}$ would contain the identity and thus coincide with $\mathfrak{A}$.

Theorem 10. *The set $\mathfrak{S}$ of all operators on an infinite-dimensional Hilbert space $\mathfrak{H}$, none having in its range a (closed) subspace of the same dimension as $\mathfrak{H}$, is an absolutely maximal non-trivial ideal in $\mathfrak{A}$.*

*Proof.* It is not difficult to verify that $\mathfrak{S}$ forms a non-trivial ideal. The preceding corollary implies that $\mathfrak{S}$ is also absolutely maximal.

In the particular case when the infinite-dimensional space is separable we have the following interesting result due to CALKIN [2, p. 841].

Theorem 11. *Let $\mathfrak{H}_0$ be a separable Hilbert space. For any ideal $\mathfrak{T}$ in $\mathfrak{A}$ we have either $\mathfrak{T} = \mathfrak{A}$ or $\mathfrak{T} \subset \mathfrak{C}$.*

*Proof.* We remark first, that an operator $A$ on $\mathfrak{H}_0$ whose range contains no infinite-dimensional (closed) subspace is necessarily completely continuous (see CALKIN [1, p. 402]). We sketch the proof of the last assertion: Since $A$ has the same range as $[A^*]$, and either both or none of the operators $A$ and $[A^*]$ is completely continuous, we may assume that $A$ is positive. But then $A$ is completely reduced by its nullspace. We may therefore also assume that $A$ is one-to-one. Let $E_\lambda$ be the resolution of the identity associated with $A$, and $\{\lambda_n\}$ be a decreasing sequence of positive numbers converging to 0 with $\lambda_1 \geqq \|A\|$, no term of which is a proper value of $A$. One then shows that the range of the projection $E_{\lambda_{n-1}} - E_{\lambda_n}$ is included in the range of $A$ and thus — by assumption — is finite-dimensional. It then follows that 0 is the only limit point of the spectrum of $A$. Theorem 9 implies that $A$ is completely continuous.

We now complete the proof. Whenever $A$ is not completely continuous, its range must contain a subspace of the same dimension as $\mathfrak{H}_0$. By Lemma 12 (Corollary), $A \in \mathfrak{T}$ implies $\mathfrak{T} = \mathfrak{A}$.

## 7. The ideals of completely continuous operators

The discussion which follows sheds additional light on the structure of ideals of completely continuous operators.

Let $A$ be completely continuous. Then $[A]$ is also completely continuous. Let $\lambda_1, \lambda_2, \ldots$ be the sequence of all the positive proper values (with repetitions) of $[A]$ arranged in a non-increasing manner. Whenever the sequence is finite, that is, of the form $\lambda_1, \ldots, \lambda_N$ we define $\lambda_{N+1} = \lambda_{N+2} = \cdots = 0$. The infinite sequence $\{\lambda_1, \lambda_2, \ldots\}$ is then defined as the *characteristic sequence* for $A$.

Let $\mathfrak{T}$ be an ideal $\subset \mathfrak{A}$. The polar decomposition for operators proves that if either of the operators $A$, $[A]$, $A^*$, $[A^*]$ is in $\mathfrak{T}$, then the remaining three are also in $\mathfrak{T}$. In particular, $\mathfrak{T}$ is self adjoint.

Assume now that an ideal $\mathfrak{T}$ contains the completely continuous operator $A$. Then every completely continuous operator $B$ with the same characteristic sequence as $A$ also belongs to $\mathfrak{T}$. This is true since $[A]$ and $[B]$ have then the same proper values. Consequently for some unitary $U$ we have $U^* [A] U = [B]$ and therefore,

$$A \in \mathfrak{T} \Leftrightarrow [A] \in \mathfrak{T} \Leftrightarrow [B] \in \mathfrak{T} \Leftrightarrow B \in \mathfrak{T}.$$

If $\mathfrak{T}$ is an ideal of completely continuous operators, then the set $\mathfrak{S}$ of the characteristic sequences of all the operators in $\mathfrak{T}$ is defined as the characteristic set of $\mathfrak{T}$. For example, the characteristic set of the ideal $\mathfrak{R}$ consists of all non-increasing sequences of non-negative numbers having only a finite number of non-zero terms.

Theorem 12. *A set $\mathfrak{S}$ of sequences $\{\lambda_i\}$ of non-negative numbers which are non-increasing and converge to* 0, *is a characteristic set of an ideal $\mathfrak{T}$ of completely continuous operators if and only if the following three conditions hold*:

(i) $\{\lambda_1, \lambda_2, \lambda_3, \ldots\} \in \mathfrak{S}, \Rightarrow \{\lambda_1, \lambda_1, \lambda_2, \lambda_2, \lambda_3, \lambda_3, \ldots\} \in \mathfrak{S}$.
(ii) $\{\lambda_i\} \in \mathfrak{S}$ *and* $\{\mu_i\} \in \mathfrak{S}, \Rightarrow \{\lambda_i + \mu_i\} \in \mathfrak{S}$.
(iii) $\{\lambda_i\} \in \mathfrak{S},\ \lambda_i \geqq \mu_i$ *and* $\mu_1 \geqq \mu_2 \geqq \cdots \geqq 0, \Rightarrow \{\mu_i\} \in \mathfrak{S}$.

*If $\mathfrak{S}$ is a characteristic set of $\mathfrak{T}$, then $\mathfrak{T}$ consists precisely of all the operators of the form $\Sigma_i \lambda_i \varphi_i \otimes \overline{\psi}_i$ where $\{\lambda_i\} \in \mathfrak{S}$ and $\{\varphi_i\}$, $\{\psi_i\}$ are any two orthonormal sequences of vectors in $\mathfrak{H}$.*

*Proof.* Let $\mathfrak{T}$ be an ideal of completely continuous operators and $\mathfrak{S}$ stand for its characteristic set. Choose a fixed orthonormal sequence $\{\varphi_i\}$ in $\mathfrak{H}$.

(i): Assume that $\{\lambda_i\} \in \mathfrak{S}$. Then $\Sigma_i \lambda_i \varphi_{2i-1} \otimes \overline{\varphi}_{2i-1}$ and $\Sigma_i \lambda_i \varphi_{2i} \otimes \overline{\varphi}_{2i}$ are in $\mathfrak{T}$. Consequently their sum is also in $\mathfrak{T}$ and thus $\{\lambda_1, \lambda_1, \lambda_2, \lambda_2, \lambda_3, \lambda_3, \ldots\}$ belongs to $\mathfrak{S}$.

(ii): $\{\lambda_i\} \in \mathfrak{S}$ and $\{\mu_i\} \in \mathfrak{S}$ implies that $\Sigma_i \lambda_i \varphi_i \otimes \overline{\varphi}_i$ and $\Sigma_i \mu_i \varphi_i \otimes \overline{\varphi}_i$ are in $\mathfrak{T}$. Hence $\Sigma_i (\lambda_i + \mu_i)\, \varphi_i \otimes \overline{\varphi}_i$ is in $\mathfrak{T}$ and thus $\{\lambda_i + \mu_i\} \in \mathfrak{S}$.

(iii): Assume that $\{\lambda_i\} \in \mathfrak{S}$ and $\lambda_i \geqq \mu_i$ as well as $\mu_1 \geqq \mu_2 \geqq \cdots \geqq 0$. Let $J$ be the set of all those subscripts $i$ for which $\mu_i > 0$. For $i \in J$ we have, $0 < \frac{\mu_i}{\lambda_i} \leqq 1$ and thus $\Sigma_{i \in J} \frac{\mu_i}{\lambda_i} \varphi_i \otimes \overline{\varphi}_i$ defines an operator. Its product with the completely continuous operator $\Sigma_i \lambda_i \varphi_i \otimes \overline{\varphi}_i$ belongs to $\mathfrak{T}$ and is obviously equal to $\Sigma_{i \in J} \mu_i \varphi_i \otimes \overline{\varphi}_i$. Thus $\{\mu_i\} \in \mathfrak{S}$.

Conversely. Let $\mathfrak{S}$ satisfy conditions (i), (ii) and (iii). Let $\mathfrak{T}$ stand for the set of all completely continuous operators of the form $\Sigma_i \lambda_i \varphi_i \otimes \overline{\psi}_i$ where $\{\lambda_i\} \in \mathfrak{S}$ and $\{\varphi_i\}$, $\{\psi_i\}$ are any two orthonormal sequences. We prove that $\mathfrak{T}$ is an ideal; $\mathfrak{S}$ is then its characteristic set. So let $A$ and $B$ belong to $\mathfrak{T}$, and $\{\lambda_i\}$, $\{\mu_i\}$ stand for their characteristic sequences. We wish to prove that $A + B$ also belongs to $\mathfrak{T}$, that is, its characteristic sequence $\{\zeta_i\}$ belongs to $\mathfrak{S}$. The $\lambda_i$'s, $\mu_i$'s and $\zeta_i$'s include of course all the positive proper values of $[A]$, $[B]$, and $[A+B]$ respectively. By Lemma 9, $\zeta_{2i} \leqq \zeta_{2i-1} \leqq 2(\lambda_i + \mu_i)$. By (ii) above, $\{\lambda_i\} \in \mathfrak{S}$ and $\{\mu_i\} \in \mathfrak{S}$ implies that $\{2\lambda_i\}$, $\{2\mu_i\}$ and hence also $\{2\lambda_i + 2\mu_i\}$ are in $\mathfrak{S}$. But then $\{2\lambda_1 + 2\mu_1, 2\lambda_1 + 2\mu_1, 2\lambda_2 + 2\mu_2, 2\lambda_2 + 2\mu_2, \ldots\}$ is also in $\mathfrak{S}$ by (i), and therefore $\{\zeta_i\} \in \mathfrak{S}$ by (iii).

Finally, let $A \in \mathfrak{T}$ and $X$ stand for any operator. Let $\{\lambda_i\}$ and $\{\mu_i\}$ stand for the characteristic sequences of $A$ and $XA$ respectively. We shall prove that $XA \in \mathfrak{T}$, that is, $\{\mu_i\} \in \mathfrak{S}$. By Lemma 8, $\mu_i \leqq \|X\| \lambda_i$. Let $N$ be a natural number $\geqq \|X\|$. By (ii), $\{\lambda_i\} \in \mathfrak{S}$ implies $\{N\lambda_i\} \in \mathfrak{S}$. Consequently, $\{\mu_i\} \in \mathfrak{S}$ by (iii). Therefore $\mathfrak{T}$ is a left ideal. Since the completely continuous operators $A$ and $A^*$ have the same characteristic sequences, the right ideal consisting of the adjoints of operators in $\mathfrak{T}$ coincides with $\mathfrak{T}$. Thus, $\mathfrak{T}$ is a two-sided ideal[1].

Corollary. *A characteristic set of a non-trivial ideal of completely continuous operators contains all non-increasing sequences of non-negative numbers having only a finite number of non-zero terms.*

## 8. Some uniformly closed left ideals of completely continuous operators

We recall that the main goal of this exposition is the study of norm ideals of operators with most of the attention centered around norm ideals of completely continuous operators. A similar study of Banach spaces (which are not necessarily norm-ideals) of completely continuous operators would change considerably the flavor of the theory even in the special case when the bound of an operator stands for its norm. While $\mathfrak{C}$ is the only uniformly closed ideal of completely continuous operators, there are many uniformly closed Banach spaces of such operators. In fact, there are many uniformly closed left ideals of completely continuous operators; the set of all completely continuos operators annihilating a given subspace of $\mathfrak{H}$ is a uniformly closed left ideal in $\mathfrak{A}$. It is true that nowhere else in this exposition do we discuss left ideals, however, the discussion which follows immediately, justifies an exception to our rule.

It is well known that any left ideal in the algebra of all linear transformations on a finite dimensional space which annihilates only the null-vector necessarily coincides with the full algebra. The generalization of the above is due to Kaplansky [1].

---

[1] The above characterization is close to the one given by Calkin [2, p. 843]. We do not assert that conditions (i), (ii) and (iii) as stated above are completely independent of each other. Their present form however, contributes to a simplification of our discussion.

Theorem 13. *A uniformly closed left ideal $\mathfrak{T}$ of completely continuous operators which annihilates only the null-vector necessarily coincides with all of $\mathfrak{C}$.*

*Proof.* Let $\mathfrak{S}$ be the set of all those vectors $\varphi \in \mathfrak{H}$ for which $\varphi \otimes \overline{\varphi}$ is in $\mathfrak{T}$.

$\mathfrak{S}$ is linear: That $\mathfrak{S}$ is closed under scalar multiplication is immediate. Now let $\varphi$ and $\psi$ be two linearly independent vectors of $\mathfrak{S}$. We prove that $\varphi + \psi$ is in $\mathfrak{S}$. Whenever $a\varphi + b\psi$ is annihilated by both $\varphi \otimes \overline{\varphi}$ and $\psi \otimes \overline{\psi}$, we have

$$(a\varphi + b\psi, \varphi) = 0 \quad \text{and} \quad (a\varphi + b\psi, \psi) = 0$$

and therefore also $a\varphi + b\psi = 0$ since

$$\|a\varphi + b\psi\|^2 = \bar{a}\,(a\varphi + b\psi, \varphi) + \bar{b}\,(a\varphi + b\psi, \psi) = 0\,.$$

Let $\mathfrak{M}$ be the two-dimensional subspace generated by $\varphi$ and $\psi$, and $\mathfrak{T}_0$ the set of those operators in $\mathfrak{T}$ which transform $\mathfrak{M}$ into itself and annihilate its orthogonal complement. Clearly, $\mathfrak{T}_0$ is a left ideal in the ring of all linear transformations on $\mathfrak{M}$, and since it contains $\varphi \otimes \overline{\varphi}$ and $\psi \otimes \overline{\psi}$ it annihilates in $\mathfrak{M}$ only the 0 vector. By the remark predecing this theorem, $\mathfrak{T}_0$ must be the full ring of linear transformations on $\mathfrak{M}$. In particular $(\varphi + \psi) \otimes \overline{(\varphi + \psi)}$ is in $\mathfrak{T}_0$, hence $\varphi + \psi$ is in $\mathfrak{S}$.

$\mathfrak{S}$ is closed: Let $\varphi_n \in \mathfrak{S}$ and $\varphi_n \to \varphi$. Then

$$\begin{aligned}\|\varphi \otimes \overline{\varphi} - \varphi_n \otimes \overline{\varphi}_n\| &\leqq \|\varphi \otimes \overline{(\varphi - \varphi_n)}\| + \|(\varphi - \varphi_n) \otimes \overline{\varphi}_n\| \\ &= \|\varphi\|\,\|\varphi - \varphi_n\| + \|\varphi_n - \varphi\|\,\|\varphi_n\| \to 0\,.\end{aligned}$$

Since $\varphi_n \otimes \overline{\varphi}_n$ is in $\mathfrak{T}$, the same is true for $\varphi \otimes \overline{\varphi}$, that is, $\varphi \in \mathfrak{S}$.

We have $\mathfrak{S} = \mathfrak{H}$: Otherwise there would be a non-zero vector $\varphi_0$ in $\mathfrak{H} \ominus \mathfrak{S}$. Since $\mathfrak{T}$ annihilates only 0, it contains an operator $A$ such that $A\varphi_0 = \psi_0 \neq 0$. Clearly,

$$A^*(\psi_0 \otimes \overline{\psi}_0)A = A^*\psi_0 \otimes \overline{A^*\psi_0}$$

is also in $\mathfrak{T}$. Consequently $A^*\psi_0$ is in $\mathfrak{S}$ and therefore $(\varphi_0, A^*\psi_0) = 0$. This contradicts

$$(\varphi_0, A^*\psi_0) = (A\varphi_0, \psi_0) = (\psi_0, \psi_0) > 0\,.$$

Thus for any $\psi \in \mathfrak{H}$, the operator $\psi \otimes \overline{\psi}$ is in $\mathfrak{T}$. Consequently,

$$\frac{1}{\|\psi\|^2}(\varphi \otimes \overline{\psi})(\psi \otimes \overline{\psi}) = \varphi \otimes \overline{\psi}$$

is also in $\mathfrak{T}$. It follows that all the operators of finite rank are in $\mathfrak{T}$ and therefore its uniform closure must be $\mathfrak{C}$.

# II. The Schmidt-class

## 1. $(\sigma c)$ as a Hilbert space of completely continuous operators

We consider here those completely continuous operators $A$ for which the squares of the positive proper values of $[A]$ form a convergent series. These may be interpreted as the integral operators on an abstract Hilbert space in the sense specified in the following section. The reader will do well to supplement the discussion of this chapter by consulting HELLINGER and TOEPLITZ [1], especially for their very illuminating bibliographic account of the theory of integral equations.

Lemma 1. *Let $A$ be an operator and $\{\varphi_j\}$, $\{\psi_i\}$ two complete orthonormal families of vectors. The families*

$$(1) \qquad \{\|A\varphi_j\|^2\}\,, \quad \{\|A^*\psi_i\|^2\}\,, \quad \{|(A\varphi_j,\, \psi_i)|^2\}$$

*of non-negative numbers are simultaneously summable or not. Whenever they are summable, their sum is the same independent of $\{\varphi_j\}$ and $\{\psi_i\}$.*

*Proof.* Parseval's equality gives

$$\|A\varphi_j\|^2 = \Sigma_i |(A\varphi_j,\, \psi_i)|^2\,.$$

Hence, if either of the above families is summable we have

$$\Sigma_j \|A\varphi_j\|^2 = \Sigma_{i,j}\,|(A\varphi_j,\, \psi_i)|^2 = \Sigma_{i,j}\,|(A^*\psi_i,\, \varphi_j)|^2 = \Sigma_i\,\|A^*\psi_i\|^2\,.$$

To complete the proof one uses the relation $A^{**} = A$.

Definition 1. We denote by $(\sigma(A))^2$ the common sum of the families (1), whenever they are summable. In the contrary case we define $\sigma(A) = +\infty$. The operators $A$ for which $\sigma(A) < +\infty$ form the *E. Schmidt-class* $(\sigma c)$.

Lemma 2. *We have*

$$\|A\| \leqq \sigma(A)\,.$$

*Proof.* The case $\sigma(A) = +\infty$ is obvious. To prove our assertion when $\sigma(A) < +\infty$, it is sufficient to show that $\|A\varphi\| \leqq \sigma(A)$ for all vectors $\varphi$ with $\|\varphi\| = 1$. That is easy. Choose a basis $\{\varphi_j\}$ with $\varphi$ as one of its elements. Then

$$\|A\varphi\| \leqq (\Sigma_j \|A\varphi_j\|^2)^{\frac{1}{2}} = \sigma(A)\,.$$

Lemma 3.

(i) $A \in (\sigma c)$ *if and only if* $A^* \in (\sigma c)$. *Moreover,* $\sigma(A) = \sigma(A^*)$.

(ii) $A \in (\sigma c)$ *implies that for any complex* $c$ also $(cA) \in (\sigma c)$ *and* $\sigma(cA) = |c|\,\sigma(A)$.

(iii) $A \in (\sigma c)$ *and* $B \in (\sigma c)$ *implies that* $(A + B) \in (\sigma c)$ *and* $\sigma(A + B) \leqq \sigma(A) + \sigma(B)$.

(iv) $A \in (\sigma c)$ *and* $X \in \mathfrak{A}$ *implies that* $AX$ *and* $XA$ *are in* $(\sigma c)$; *moreover both* $\sigma(AX)$ *and* $\sigma(XA)$ *are* $\leqq \|X\| \sigma(A)$.

(v) *For any pair of vectors* $\varphi$ *and* $\psi$, *the operator* $(\varphi \otimes \overline{\psi}) \in (\sigma c)$ *and* $\sigma(\varphi \otimes \overline{\psi}) = \|\varphi\| \|\psi\|$.

(vi) $A \in (\sigma c)$ *if and only if* $[A] \in (\sigma c)$. *Moreover,* $\sigma(A) = \sigma([A])$.

*Proof.* (i) and (ii) are immediate.

(iii): Choose a basis $\{\varphi_j\}$. It is then sufficient to observe that for any finite set $J$ of indices $j$ we have

$$(\Sigma_{j \in J} \|(A + B)\varphi_j\|^2)^{\frac{1}{2}} \leqq (\Sigma_{j \in J}(\|A\varphi_j\| + \|B\varphi_j\|)^2)^{\frac{1}{2}} \leqq$$
$$\leqq (\Sigma_{j \in J} \|A\varphi_j\|^2)^{\frac{1}{2}} + (\Sigma_{j \in J} \|B\varphi_j\|^2)^{\frac{1}{2}} \leqq \sigma(A) + \sigma(B) .$$

(iv): Since $\|XA\varphi_j\|^2 \leqq \|X\|^2 \|A\varphi_j\|^2$ the first sum of Lemma 1 gives $\sigma(XA) \leqq \|X\| \sigma(A)$. Using the last inequality and (i) above, one concludes,

$$\sigma(AX) = \sigma((AX)^*) = \sigma(X^*A^*) \leqq \|X^*\| \sigma(A^*) = \|X\| \sigma(A) .$$

(v): Choose a basis $\{\varphi_j\}$. By Parseval's equality

$$\|\varphi\|^2 \|\psi\|^2 = \|\varphi\|^2 \Sigma_j |(\varphi_j, \psi)|^2 = \Sigma_j \|(\varphi_j, \psi)\varphi\|^2 = \Sigma_j \|(\varphi \otimes \overline{\psi})\varphi_j\|^2 .$$

(vi): We use the polar decomposition $A = W[A]$, $[A] = W^*A$ and apply (iv).

Remark. It is a consequence of (v) and (iii) that all operators of finite rank belong to $(\sigma c)$.

Remark. We shall prove later that in the particular case of the space $L^2$, the operators in $(\sigma c)$ correspond precisely to the integral operators on $L^2$ (whose kernels are elements of $\boldsymbol{L}^2$). For integral operators on $L^2$, statements (i)—(vi) are well known and frequently used in analysis. For instance, (iv) asserts that the product of two operators of which one is an integral operator is also an integral operator.

Lemma 4. *Let* $A$ *and* $B$ *belong to* $(\sigma c)$ *and* $\{\varphi_j\}$ *be a basis. The family* $\{|(A\varphi_j, B\varphi_j)|\}$ *is then summable. Consequently, the family* $\{(A\varphi_j, B\varphi_j)\}$ *is also summable; its sum is independent of* $\{\varphi_j\}$.

*Proof.* Clearly, $|(A\varphi_j, B\varphi_j)| \leqq \frac{1}{2}(\|A\varphi_j\|^2 + \|B\varphi_j\|^2)$ and therefore,

$$\Sigma_j |(A\varphi_j, B\varphi_j)| \leqq \tfrac{1}{2}((\sigma(A))^2 + (\sigma(B))^2) .$$

Since,

$$\mathscr{R}(A\varphi_j, B\varphi_j) = \tfrac{1}{4}(\|(A + B)\varphi_j\|^2 - \|(A - B)\varphi_j\|^2) ,$$

we also have,

$$\mathscr{R}\Sigma_j(A\varphi_j, B\varphi_j) = \tfrac{1}{4}((\sigma(A + B))^2 - (\sigma(A - B))^2)$$

and the right side is clearly independent of $\{\varphi_j\}$. Replacing $A$ by $iA$, we see that

$$\mathscr{I}\Sigma_j(A\varphi_j, B\varphi_j) = -\mathscr{R}\Sigma_j(iA\varphi_j, B\varphi_j)$$

is also independent of $\{\varphi_j\}$.

Definition 2. Let $A$ and $B$ belong to ($\sigma$ $c$) and $\{\varphi_j\}$ be a basis in $\mathfrak{H}$. Define

$$(A, B) = \Sigma_j (A\varphi_j, B\varphi_j) .$$

The preceding lemma states that $(A, B)$ *is independent of the chosen basis.*

In the lemma which follows, the operators $A$, $B$, $A_i$, $B_i$ are in ($\sigma$ $c$), $X$ is an arbitrary operator, $c$ is a complex number.

Lemma 5.

(i) $(B, A) = \overline{(A, B)}$.
(ii) $(cA, B) = c(A, B)$.
(ii') $(A, cB) = \bar{c}(A, B)$.
(iii) $(A_1 + A_2, B) = (A_1, B) + (A_2, B)$.
(iii') $(A, B_1 + B_2) = (A, B_1) + (A, B_2)$.
(iv) $(A, A) \geqq 0$; $(A, A) = 0$ *only for* $A = 0$.
(v) $(A^*, B^*) = \overline{(A, B)}$.
(vi) $(XA, B) = (A, X^*B)$.
(vi') $(AX, B) = (A, BX^*)$.
(vii) *For any four vectors* $\varphi$, $\psi$, $f$ *and* $g$ *we have* $(\varphi \otimes \overline{\psi}, f \otimes \bar{g}) = (\varphi, f)(g, \psi)$.

*Proof.* (i) through (iii') are immediate consequences of Definition 2.

(iv): If $A \neq 0$, then for some $\varphi$ of norm 1, $A\varphi \neq 0$. Since we may obviously choose a basis with $\varphi$ as one of its elements, we have

$$(A, A) \geqq (A\varphi, A\varphi) = \|A\varphi\|^2 > 0 .$$

That $A = 0$ implies $(A, A) = 0$ is obvious.

(v): As in the proof of Lemma 4, observe that

$$\begin{aligned}\mathcal{R}(A, B) &= \tfrac{1}{4}\left((\sigma(A + B))^2 - (\sigma(A - B))^2\right) \\ &= \tfrac{1}{4}\left((\sigma(A^* + B^*))^2 - (\sigma(A^* - B^*))^2\right) = \mathcal{R}(A^*, B^*) .\end{aligned}$$

Replacing $A$ by $iA$ one gets

$$\mathcal{I}(A, B) = -\mathcal{R}(iA, B) = -\mathcal{R}(-iA^*, B^*) = -\mathcal{I}(A^*, B^*) .$$

(vi): Observe first that both $XA$ and $X^*B$ belong to ($\sigma$ $c$). The desired conclusion follows from $(XA\varphi_j, B\varphi_j) = (A\varphi_j, X^*B\varphi_j)$.

(vi'): Using (v) and (vi) one gets

$$(AX, B) = \overline{(X^*A^*, B^*)} = \overline{(A^*, XB^*)} = (A, BX^*) .$$

(vii): Choose a basis $\{\varphi_j\}$. Then

$$\Sigma_j((\varphi_j, \psi)\varphi, (\varphi_j, g) f) = (\varphi, f)\, \Sigma_j (g, \varphi_j)(\varphi_j, \psi) = (\varphi, f)(g, \psi) ,$$

where the last equality follows from Parseval's identity.

Remark. In the particular case when $\varphi_1, \ldots, \varphi_n$; $\psi_1, \ldots, \psi_n$ and $f_1, \ldots, f_m$; $g_1, \ldots, g_m$ are arbitrary vectors, (iii), (iii') and (vii) implies that

$$(\Sigma_{i=1}^n \varphi_i \otimes \overline{\psi}_i, \Sigma_{j=1}^m f_j \otimes \bar{g}_j) = \Sigma_{i=1}^n \Sigma_{j=1}^m (\varphi_i, f_j)(g_j, \psi_i) .$$

Statements (ii) and (iii) of Lemma 3 imply that $(\sigma c)$ is a linear space, while (i)—(iv) of Lemma 5 state that $(A, B)$ defines there an inner product and that

$$(A, A)^{\frac{1}{2}} = \sigma(A)$$

is the norm that goes with it. In particular, we have *Schwarz's inequality*

$$|(A, B)| \leqq \sigma(A)\,\sigma(B)\,;$$

the equality sign holds if and only $A$ and $B$ are linearly dependent.

Lemma 6. *The Schmidt-class of operators is a complete space with respect to its metric* $\sigma(A - B)$.

*Proof.* With a slight modification one follows the usual argument in the proof of completeness of the sequential Hilbert space (consisting of all sequences $\{\zeta_i\}$ of complex numbers such that $\Sigma_i |\zeta_i|^2 < +\infty$). Assume thus

$$A_n \in (\sigma c) \quad \text{and} \quad \sigma(A_n - A_m) \to 0 \quad \text{as} \quad m, n \to +\infty.$$

Since

$$\|A_n - A_m\| \leqq \sigma(A_n - A_m)\,,$$

the sequence $\{A_n\}$ converges uniformly, and thus it must converge uniformly to some operator $A$. We prove that

$$A \in (\sigma c) \quad \text{and} \quad \sigma(A - A_n) \to 0\,.$$

Given $\varepsilon > 0$, choose $N$ so that

$$\sigma(A_n - A_m) < \varepsilon \quad \text{for} \quad n, m \geqq N\,.$$

This of course implies (when a fixed basis $\{\varphi_j\}$ is chosen)

$$\Sigma_{j \in J} \|(A_n - A_m)\varphi_j\|^2 < \varepsilon^2$$

for $m, n \geqq N$ and any finite set $J$ of indices $j$. Thus,

$$\Sigma_{j \in J} \|(A_n - A)\varphi_j\|^2 \leqq \varepsilon^2$$

for $n \geqq N$ and any finite $J$. Consequently, $(A_n - A) \in (\sigma c)$ and therefore also $A \in (\sigma c)$. Moreover,

$$\sigma(A - A_n) \leqq \varepsilon \quad \text{for all} \quad n \geqq N\,.$$

Theorem 1. *Every operator $A$ in $(\sigma c)$ is necessarily completely continuous.*

*Proof.* Let $f_n \rightharpoonup f$. We prove the following: Given an $\varepsilon > 0$, there is a natural number $n_0$ such that for all $n > n_0$ one has $\|Af_n - Af\| < \varepsilon$.

Clearly for some constant $c$ we have $\|f_n\| \leqq c$ for all $n$ and thus also $\|f\| \leqq c$. Choose a basis $\{\varphi_j\}$. Since $A \in (\sigma c)$, we can find a finite set $J$ of indices $j$ such that

$$\Sigma_{j \notin J} \|A\,\varphi_j\|^2 < \frac{\varepsilon^2}{16\,c^2}\,.$$

We have,

$$f_n - f = \Sigma_j (f_n - f, \varphi_j)\varphi_j$$

and therefore,

$$Af_n - Af = \Sigma_j (f_n - f, \varphi_j) A\varphi_j .$$

Thus, for every natural $n$,

$$\|Af_n - Af\|^2 = \|\Sigma_{j\in J} (f_n - f, \varphi_j) A\varphi_j + \Sigma_{j\notin J} (f_n - f, \varphi_j) A\varphi_j\|^2 \leqq$$
$$\leqq 2\|\Sigma_{j\in J} (f_n - f, \varphi_j) A\varphi_j\|^2 + 2\|\Sigma_{j\notin J} (f_n - f, \varphi_j) A\varphi_j\|^2 .$$

The second term on the right of the last inequality is

$$\leqq 2\,(\Sigma_{j\notin J} |(f_n - f, \varphi_j)|\, \|A\varphi_j\|)^2 \leqq$$
$$\leqq 2\,\Sigma_{j\notin J} |(f_n - f, \varphi_j)|^2\, \Sigma_{j\notin J} \|A\varphi_j\|^2 \leqq$$
$$\leqq 2\,\|f_n - f\|^2\, \Sigma_{j\notin J} \|A\varphi_j\|^2 \leqq 2(2c)^2 \frac{\varepsilon^2}{16\,c^2} = \frac{\varepsilon^2}{2} .$$

On the other hand $f_n \rightharpoonup f$ implies $\lim\limits_n (f_n - f, \varphi_j) = 0$ for all $j$. Consequently, the expression

$$2\|\Sigma_{j\in J} (f_n - f, \varphi_j) A\varphi_j\|^2$$

involving a sum with a finite number of terms can be made $< \frac{\varepsilon^2}{2}$ for large enough $n$, say for all $n > n_0$. Therefore for all $n > n_0$, we have $\|Af_n - Af\|^2 < \varepsilon^2$.

Remark. We have proved already for the special case when $\mathfrak{H}$ is separable, that any proper ideal of operators is necessarily included in $\mathfrak{C}$. That ($\sigma\, c$) is an ideal follows of course from Lemma 3.

Theorem 2. *A completely continuous operator* $A = \Sigma_i \lambda_i \varphi_i \otimes \overline{\psi}_i$ *belongs to* ($\sigma\, c$) *if and only if* $\Sigma_i \lambda_i^2 < +\infty$, *that is, the series formed from the non-zero proper values of* $A^*A$ *converges. In the last case we also have*

$$\sigma(A) = (\Sigma_i \lambda_i^2)^{\frac{1}{2}} .$$

*Proof.* We extend $\{\psi_i\}$ to a complete orthonormal family by adding $\{\omega_j\}$. Of course

$$\|A\psi_i\| = \|\lambda_i \varphi_i\| = \lambda_i \quad \text{and} \quad \|A\omega_j\| = 0 .$$

Hence,

$$(\sigma(A))^2 = \Sigma_i \|A\psi_i\|^2 + \Sigma_j \|A\omega_j\|^2 = \Sigma_i \lambda_i^2 .$$

Corollary. *The operators of finite rank form a dense set in* ($\sigma\, c$).

*Proof.* Let $A \in (\sigma\, c)$. Being completely continuous $A$ may be written in the polar form $A = \Sigma_i \lambda_i \varphi_i \otimes \overline{\psi}_i$ with $\Sigma_i \lambda_i^2 < +\infty$. Let $A_n = \Sigma_{i=1}^n \lambda_i \varphi_i \otimes \overline{\psi}_i$. Then,

$$\lim_n \sigma(A - A_n) = \lim_n (\Sigma_{i>n} \lambda_i^2)^{\frac{1}{2}} = 0 .$$

We summarize the preceding discussion:

Theorem 3. *Let* $\{\varphi_j\}$ *be a basis in* $\mathfrak{H}$. *The set* $(\sigma\, c)$ *of all operators* $A$ *for which* $\{\|A\varphi_j\|^2\}$ *is summable, is a linear space. There,* $(\Sigma_j\|A\varphi_j\|^2)^{\frac{1}{2}} = \sigma(A)$ *is a norm. The resulting normed linear space is complete hence a Banach space; it contains the operators of finite rank as a dense subset. For any pair of operators* $A$ *and* $B$ *in* $(\sigma\, c)$, *the family* $\{(A\varphi_j, B\varphi_j)\}$ *of complex numbers is summable. Its sum* $(A, B)$ *defines an inner product in* $(\sigma\, c)$ *and* $(A, A)^{\frac{1}{2}} = \sigma(A)$ *is the norm that goes with it. Thus,* $(\sigma\, c)$ *is a Hilbert space (independent on the chosen basis* $\{\varphi_j\}$*). The operators in* $(\sigma\, c)$ *are necessarily completely continuous. They are precisely those completely continuous operators (in the polar form)* $A = \Sigma_i \lambda_i \omega_i \otimes \overline{\psi}_i$ *for which* $\Sigma_i \lambda_i^2 < +\infty$; *we also have* $\sigma(A) = (\Sigma_i \lambda_i^2)^{\frac{1}{2}}$. *Moreover,* $(\sigma\, c)$ *is an ideal in the algebra* $\mathfrak{A}$. *Under its own norm* $(\sigma\, c)$ *is a Banach algebra, in fact, also a norm-ideal.*

## 2. The Schmidt-class of operators on $L^2$

We consider $L^2$ and $\boldsymbol{L}^2$. To avoid confusion we find it best to denote the inner product and its corresponding norm in $L^2$ by $(\,,)$ and $\|\;\|$. Similarly, $(\,,)$ and $|\;|$ will stand for the inner product and norm in $\boldsymbol{L}^2$.

Let $\mathscr{K}(x, y)$ be a fixed element in $\boldsymbol{L}^2$. For $f(x)$ in $L^2$, the integral

$$\int \mathscr{K}(x, y) f(y)\, dy$$

is defined for almost all $x$, and represents a function $g(x)$ also in $L^2$. In fact, the equation

$$g(x) = \int \mathscr{K}(x, y) f(y)\, dy \tag{1}$$

defines an operator $K$ on $L^2$, which for obvious reasons will be termed an *integral operator* or of integral type. It is a consequence of Fubini's theorem on successive integration and Schwarz's inequality that

$$\|K\| \leqq |\mathscr{K}|.$$

One observes that there are many functions $\mathscr{K}(x, y)$ not necessarily belonging to $\boldsymbol{L}^2$ for which the last equality will furnish a well defined transformation (perhaps unbounded) on $L^2$ (see Stone [1, p. 101]). Such functions will be of no interest in the present discussion. For us an integral operator $K$ on $L^2$, will always mean one of the type (1) generated by a function $\mathscr{K}(x, y)$ in $\boldsymbol{L}^2$.

One recalls further that if $K_0$, $K_1$, $K_2$ are integral operators on $L^2$ generated by the kernels $\mathscr{K}_0(x, y)$, $\mathscr{K}_1(x, y)$, $\mathscr{K}_2(x, y)$ in $\boldsymbol{L}^2$, then $K_0^*$, $cK_0$, $K_1 + K_2$ are also integral operators generated by the kernels $\overline{\mathscr{K}_0(y, x)}$, $c\mathscr{K}_0(x, y)$, $\mathscr{K}_1(x, y) + \mathscr{K}_2(x, y)$ respectively. Thus, $K_0$ is Hermitean if and only if its generating kernel $\mathscr{K}_0(x, y)$ is Hermitean; we mean by this that $\mathscr{K}_0(x, y) = \overline{\mathscr{K}_0(y, x)}$ (almost everywhere).

Moreover, $K_1 K_2$ is an integral operator generated by the function

(2) $$\mathscr{H}(x, y) = \int \mathscr{K}_1(x, z)\, \mathscr{K}_2(z, y)\, dz$$

which also belongs to $\mathbf{L}^2$. We also have

$$|\mathscr{H}| \leqq |\mathscr{K}_1|\, |\mathscr{K}_2| .$$

Theorem 4. *An operator $K$ on $L^2$ is of integral type if and only if it belongs to the Schmidt-class. Whenever $\mathscr{K}(x, y)$ in $\mathbf{L}^2$ generates $K$, then*

$$\sigma(K) = |\mathscr{K}| .$$

*The correspondence $K \leftrightarrow \mathscr{K}$ thus establishes an equivalence relation between $\mathbf{L}^2$ and the Schmidt-class of operators on $L^2$. Moreover, if $\mathscr{H}$ defined by (2) above is considered as the "product" of $\mathscr{K}_1$ and $\mathscr{K}_2$ in $\mathbf{L}^2$, one may say that $\mathbf{L}^2$ and the Schmidt-class of operators on $L^2$ are also equivalent as Banach algebras.*

*Proof.* Let $K$ be an integral operator on $L^2$ generated by $\mathscr{K} \in \mathbf{L}^2$. If $\varphi(x)$ and $\psi(x)$ are in $L^2$, then obviously $\psi(x)\,\overline{\varphi(y)}$ is in $\mathbf{L}^2$ and

$$\begin{aligned}(K\varphi, \psi) &= \int \left(\int \mathscr{K}(x, y)\, \varphi(y)\, dy\right) \overline{\psi(x)}\, dx \\ &= \iint \mathscr{K}(x, y)\, \varphi(y)\, \overline{\psi(x)}\, dx\, dy = (\mathscr{K}(x, y), \psi(x)\, \overline{\varphi(y)}) .\end{aligned}$$

Now choose any two complete orthonormal sequences $\{\varphi_i\}$, $\{\psi_j\}$ in $L^2$ (in particular they may be the same). Then $\{\psi_j(x)\, \overline{\varphi_i(y)}\}\; i, j = 1, 2, \ldots$ is known to be a complete orthonormal sequence in $\mathbf{L}^2$. An application of Parseval's equality gives

$$\Sigma_{i,j}\, |(K\varphi_i, \psi_j)|^2 = \Sigma_{i,j}\, |(\mathscr{K}(x, y), \psi_j(x)\, \overline{\varphi_i(y)})|^2 = |\mathscr{K}|^2 .$$

This proves so far that an integral operator $K$ on $L^2$ generated by $\mathscr{K}(x, y)$ in $\mathbf{L}^2$, belongs to the Schmidt-class and $\sigma(K) = |\mathscr{K}|$.

To prove the converse we remark first that every operator on $L^2$ of finite rank $\Sigma_{i=1}^{n} \lambda_i\, \varphi_i \otimes \overline{\psi}_i$ is an integral operator corresponding to the degenerate kernel $\Sigma_{i=1}^{n} \lambda_i\, \varphi_i(x)\, \overline{\psi_i(y)}$. This is precisely the reason why the operator in Definition I.1 was denoted by $\varphi \otimes \overline{\psi}$ rather than $\varphi \otimes \psi$.

Now, an operator $A$ in $(\sigma c)$ is completely continuous and thus admits the polar representation $A = \Sigma_i \lambda_i\, \varphi_i \otimes \overline{\psi}_i$ with $\Sigma_i \lambda_i^2 < +\infty$. Denoting by $\mathscr{K}_n$ the degenerate kernel in $\mathbf{L}^2$ corresponding to the integral operator $K_n = \Sigma_{i=1}^{n} \lambda_i\, \varphi_i \otimes \overline{\psi}_i$ we have for $n > m$

$$\lim_{m,n} |\mathscr{K}_n - \mathscr{K}_m| = \lim_{m,n} \sigma(K_n - K_m) = \lim_{m,n} (\Sigma_{i=m+1}^{n} \lambda_i^2)^{\frac{1}{2}} = 0 .$$

The sequence $\{\mathscr{K}_n\}$ is thus fundamental in $\mathbf{L}^2$. Since the last space is known to be complete, there is a $\mathscr{K}$ in $\mathbf{L}^2$ to which our sequence converges. By what was already proven the integral operator $K$ with the kernel $\mathscr{K}$ belongs to the Schmidt-class. We also have,

$$\sigma(K - K_n) = |\mathscr{K} - \mathscr{K}_n| \to 0 .$$

On the other hand

$$\sigma(A - K_n) = (\Sigma_{i>n} \lambda_i^2)^{\frac{1}{2}} \to 0 .$$

Consequently, $A = K$, that is, $A$ is of integral type and is generated by $\mathcal{K}$.

Remark. In the light of the last theorem the operators in ($\sigma c$) play the rôle of integral operators on an abstract Hilbert space. For the special case of $L^2$, both notions coincide. While one carries over to an abstract Hilbert space without difficulty, the second does not yield to a straightforward generalization.

Corollary. *Every Hermitean $\mathcal{K}(x, y)$ in $\mathbf{L}^2$ is representable in the form*

$$\mathcal{K}(x, y) = \Sigma_i \lambda_i \varphi_i(x) \overline{\varphi_i(y)}$$

*where the $\lambda_i$'s are the non-zero proper values of the integral operator it determines on $L^2$ and $\{\varphi_i(x)\}$ is a corresponding orthonormal sequence of proper vectors; the convergence of the series on the right is that in the sense of* $\mathbf{L}^2$ (see, E. SCHMIDT [3]).

*Proof.* A Hermitean $\mathcal{K}(x, y)$ in $\mathbf{L}^2$ generates an integral operator $K$ on $L^2$ which is Hermitean and in ($\sigma c$). If $K = \Sigma_i \lambda_i \varphi_i \otimes \overline{\varphi}_i$ is its spectral form, then $\Sigma_i \lambda_i^2 < +\infty$ and

$$|\mathcal{K}(x, y) - \Sigma_{i=1}^n \lambda_i \varphi_i(x) \overline{\varphi_i(y)}| = \sigma(K - \Sigma_{i=1}^n \lambda_i \varphi_i \otimes \overline{\varphi}_i)$$
$$= \sigma(\Sigma_{i>n} \lambda_i \varphi_i \otimes \overline{\varphi}_i) = (\Sigma_{i>n} \lambda_i^2)^{\frac{1}{2}} \to 0 .$$

Analogously using the polar representation for (arbitrary) completely continuous operators we prove: *A given $\mathcal{K}(x, y)$ in $\mathbf{L}^2$ is always representable in the form*

$$\mathcal{K}(x, y) = \Sigma_i \lambda_i \varphi_i(x) \overline{\psi_i(y)}$$

*where all the $\lambda_i$'s are $> 0$ and $\{\varphi_i(x)\}$ as well as $\{\psi_i(x)\}$ form orthonormal sequences in $L^2$; the convergence of the series on the right is that in the sense of $\mathbf{L}^2$. Moreover, this representation is unique in the sense that the $\lambda_i$'s are necessarily the positive proper values of $[K]$, where $K$ is the integral operator generated by $\mathcal{K}(x, y)$.*

## III. The trace-class ($\tau c$)

### 1. ($\tau c$) as a Banach space of completely continuous operators

Consider $\mathfrak{R}_n$ — the linear space of all linear transformations on an $n$-dimensional space. The trace $t(A)$ defined for $A \in \mathfrak{R}_n$ is a complex valued linear functional whose characteristic properties are simple and well known. Moreover $\tau(A) = t([A])$ defines a norm on $\mathfrak{R}_n$. An extension of these facts is being presented in the discussion which follows.

We recall that the product of two operators of which at least one is in $(\sigma c)$ is also in $(\sigma c)$. Our present interest will center however on products of operators both of which are in $(\sigma c)$. It is precisely for these operators $A$, that we define $t(A)$ and $\tau(A)$.

Lemma 1. *Let $A$ be the product of two operators in $(\sigma c)$ and $\{\varphi_j\}$ a given basis. Then the family $\{|(A\varphi_j, \varphi_j)|\}$ of non-negative numbers is summable. Consequently, $\{(A\varphi_j, \varphi_j)\}$ is also summable and its sum is independent of $\{\varphi_j\}$.*

*Proof.* Since $C \in (\sigma c)$ if and only if $C^* \in (\sigma c)$ we may assume that $A = C^*B$ with both $B$ and $C$ in $(\sigma c)$. Then,

$$(A\varphi_j, \varphi_j) = (B\varphi_j, C\varphi_j) .$$

An application of Lemma II.4 concludes the proof.

Definition 1. The products of two operators in $(\sigma c)$ form the *trace-class* $(\tau c)$. For $A \in (\tau c)$ the finite number

$$t(A) = \Sigma_j (A\varphi_j, \varphi_j)$$

defines the *trace* of $A$.

Remark. If $B$ and $C$ are in $(\sigma c)$ and $A = C^*B$ then

$$t(A) = (B, C) .$$

Thus, if also $B_1$ and $C_1$ are in $(\sigma c)$ then

$$C^*B = C_1^*B_1 \quad \text{implies} \quad (B, C) = (B_1, C_1) .$$

Consider an operator $A$. Then $[A]$ is well defined. Since $[A] \geqq 0$, the operator $[A]^{\frac{1}{2}}$ is also well defined. The complete orthonormal family $\{\varphi_j\}$ in the lemma which follows should be viewed as arbitrarily given, but fixed, that is, the criterion is valid if applied to one basis $\{\varphi_j\}$ and it does not matter how this basis is chosen.

Lemma 2. *The following statements are equivalent:*

(i) $A \in (\tau c)$.
(ii) $[A] \in (\tau c)$.
(iii) $[A]^{\frac{1}{2}} \in (\sigma c)$.
(iv) $\Sigma_j ([A]\varphi_j, \varphi_j) < +\infty$.

*Proof.* Let $A = W[A]$, $[A] = W^*A$ be the polar decomposition of $A$. We prove: (iii) → (i) → (ii) → (iv) → (iii).

Assume (iii). Then also $W[A]^{\frac{1}{2}}$ is in $(\sigma c)$ and thus $A$ being the product of $W[A]^{\frac{1}{2}}$ and $[A]^{\frac{1}{2}}$ — both in $(\sigma c)$ — is by definition in $(\tau c)$.

Assume (i). Thus $A = BC$ with both $B$ and $C$ in $(\sigma c)$. Since $W^*B$ is also in $(\sigma c)$

$$[A] = W^*A = W^*B \cdot C$$

is in $(\tau c)$.

Assume (ii). Lemma 1 implies that (iv) holds.

Finally, (iv) is equivalent to (iii) since,

$$([A]\varphi_j, \varphi_j) = \|[A]^{\frac{1}{2}}\varphi_j\|^2.$$

Lemma 3.

(i) $A \in (\tau c)$ *if and only if* $A^* \in (\tau c)$.

(ii) $A \in (\tau c)$ *implies* $(cA) \in (\tau c)$ *for any complex* $c$.

(iii) $A$ *and* $B$ *in* $(\tau c)$ *implies that* $(A + B) \in (\tau c)$.

(iv) $A \in (\tau c)$ *and* $X \in \mathfrak{A}$ *implies that* $AX$ *and* $XA$ *are in* $(\tau c)$.

*Proof.* (i) is true since $A = BC$ is equivalent to $A^* = C^*B^*$ and $B \in (\sigma c)$ if and only if $B^* \in (\sigma c)$.

(ii) is immediate.

(iv). Let $A = BC$ with $B$ and $C$ in $(\sigma c)$. Then $CX$ and $XB$ are also in$(\sigma c)$ and $AX = B(CX)$, $XA = (XB)\,C$.

(iii). We use the polar decomposition for $(A + B)$. Then

$$[A + B] = W^*(A + B) = W^*A + W^*B\,.$$

By (iv) above, both $W^*A$ and $W^*B$ are in $(\tau c)$ and thus, $\Sigma_j(W^*A\varphi_j, \varphi_j)$ and $\Sigma_j(W^*B\varphi_j, \varphi_j)$ are well defined. Since

$$([A + B]\varphi_j, \varphi_j) = (W^*A\varphi_j, \varphi_j) + (W^*B\varphi_j, \varphi_j)$$

we have

$$\Sigma_j\,([A + B]\varphi_j, \varphi_j) < +\infty\,.$$

By Lemma 2, this is equivalent to $(A + B) \in (\tau c)$.

In (i)—(iii) below, both $A$ and $B$ are in $(\tau c)$; in (iv) either $A$ or $B$ is in $(\tau c)$. The expressions in (i)—(iv) are defined by Lemma 3.

Lemma 4.

(i) $t(A^*) = \overline{t(A)}$.

(ii) $t(cA) = c\,t(A)$.

(iii) $t(A + B) = t(A) + t(B)$.

(iv) $t(AB) = t(BA)$.

*Proof.* (i)—(iii) is obvious.

(iv): Assume $A \in (\tau c)$ and put $X$ for $B$. Let also $A = C^*B$ where both $B$ and $C$ in $(\sigma c)$. We have,

$$t(AX) = t(C^*BX) = (BX, C)\,,$$

$$t(XA) = t(XC^*B) = (B, CX^*)\,.$$

The two expressions on the extreme right are equal to each other by Lemma II.5. (vi′).

Definition 2. For $A \in (\tau c)$, define

$$\tau(A) = t([A])\,.$$

It is a consequence of Lemma 2 that $\tau(A)$ is a well defined finite non-negative number for every operator $A$ in $(\tau c)$. Whenever $A$ is not in $(\tau c)$ put $\tau(A) = +\infty$.

Lemma 5. *$A \in (\tau c)$ if and only if $[A] \in (\tau c)$. We have,*

$$\tau(A) = \tau([A]) .$$

*Proof.* In view of Lemma 2 it remains to prove the last equality. Since,

$$[[A]] = [A]$$

we have,

$$\tau(A) = t([A]) = t([[A]]) = \tau([A]) .$$

Lemma 6. *$A \in (\tau c)$ if and only if $[A]^{\frac{1}{2}} \in (\sigma c)$. We have,*

$$\tau(A) = (\sigma([A]^{\frac{1}{2}}))^2 .$$

*Proof.* It remains to prove the last equality:

$$\tau(A) = t([A]) = ([A]^{\frac{1}{2}}, [A]^{\frac{1}{2}}) = (\sigma([A]^{\frac{1}{2}}))^2 .$$

Lemma 7. *Let $A \in (\tau c)$ and $X$ be any operator. Then $X[A]$ is in $(\tau c)$ and*

$$|t(X[A])| \leqq \|X\| \, \tau(A) .$$

*Proof.* It remains to prove the last inequality. With the previous lemma in mind one reasons as follows: Clearly, $[A]^{\frac{1}{2}}$ and thus also $X[A]^{\frac{1}{2}}$ and $[A]^{\frac{1}{2}}X^*$ are in $(\sigma c)$. Moreover,

$$t(X[A]) = t(X[A]^{\frac{1}{2}}[A]^{\frac{1}{2}}) = ([A]^{\frac{1}{2}}, [A]^{\frac{1}{2}}X^*) .$$

Applying Schwarz's inequality one gets

$$|t(X[A])| \leqq \sigma([A]^{\frac{1}{2}}) \, \sigma([A]^{\frac{1}{2}}X^*) .$$

The right side is clearly

$$\leqq \|X\| \, (\sigma([A]^{\frac{1}{2}}))^2 = \|X\| \, \tau(A) .$$

Lemma 8. *In what follows, $A$ and $B$ are in $(\tau c)$, $X$ is in $\mathfrak{A}$, $c$ is complex.*

(i) $\tau(A^*) = \tau(A)$.
(ii) $\tau(cA) = |c| \, \tau(A)$.
(iii) $\tau(A + B) \leqq \tau(A) + \tau(B)$.
(iv) $\tau(A) \geqq 0$; $\tau(A) = 0$ *implies* $A = 0$.
(v) *Both* $\tau(AX)$ *and* $\tau(XA)$ *are* $\leqq \|X\| \, \tau(A)$.
(vi) $|t(A)| \leqq \tau(A)$.

*Proof.* (i): Let $W$ be the partially isometric operator determined by the polar decomposition of $A$. One has

$$\tau(A^*) = t([A^*]) = t(W[A]W^*) = t(W^*W[A]) = t([A]) = \tau(A) .$$

(ii): Obvious, since $(cA)^*(cA) = |c|^2 A^*A$ and thus

$$[cA] = |c|\,[A] .$$

(v): Let $W$ and $W_1$ be the partially isometric operators determined by the polar decomposition of $A$ and $XA$. Then

$$A = W[A] \quad \text{and} \quad [XA] = W_1{}^*XA$$

and therefore $[XA] = Y[A]$, where $Y = W_1{}^*XW$ and thus $\|Y\| \leqq \|X\|$. Now,

$$\tau(XA) = t([XA]) = t(Y[A]) \leqq \|Y\|\,\tau(A) \leqq \|X\|\,\tau(A);$$

the first inequality sign being justified by Lemma 7. Using this result and (i) above, one obtains a similar inequality for $AX$.

(vi): Applying Lemma 7 to $A = W[A]$ one has

$$|t(A)| = |t(W[A])| \leqq \|W\|\,\tau(A) \leqq \tau(A) .$$

(iii): Determining the partially isometric $W$, $W_1$, $W_2$ so that

$$A = W[A], \quad B = W_1[B], \quad [A+B] = W_2^*(A+B)$$

one has

$$[A+B] = W_2^*W[A] + W_2^*W_1[B] .$$

Hence,

$$\tau(A+B) = t([A+B]) = t(W_2^*W[A]) + t(W_2^*W_1[B]) .$$

By Lemma 7, the extreme right of the last equality is

$$\leqq \|W_2^*W\|\,\tau(A) + \|W_2^*W_1\|\,\tau(B) \leqq \tau(A) + \tau(B) .$$

(iv): Since $\tau(A) = (\sigma([A]^{\frac{1}{2}}))^2$ it is clear that $\tau(A) \geqq 0$. Now,

$$\tau(A) = 0 \Leftrightarrow [A]^{\frac{1}{2}} = 0 \Leftrightarrow [A]^2 = A^*A = 0 \Leftrightarrow A = 0 .$$

Corollary. *For $A \in (\tau c)$ and two bases $\{\varphi_j\}$, $\{\psi_j\}$, the family $\{|(A\varphi_j, \psi_j)|\}$ is summable. Hence $\{(A\varphi_j, \psi_j)\}$ is also summable. Moreover,*

$$|\Sigma_j\,(A\varphi_j, \psi_j)| \leqq \tau(A) .$$

*Proof.* Let $U$ be the unitary operator transforming $\varphi_j$ into $\psi_j$. Then, $(A\varphi_j, \psi_j) = (A\varphi_j, U\varphi_j) = (U^*A\varphi_j, \varphi_j)$. Since, $U^*A$ is also in $(\tau c)$, Lemma 1 implies the summability of $\{|(A\varphi_j, \psi_j)|\}$. Furthermore,

$$|\Sigma_j\,(A\varphi_j, \psi_j)| = |\Sigma_j(U^*A\varphi_j, \varphi_j)| = |t(U^*A)| \leqq$$
$$\leqq \tau(U^*A) \leqq \|U^*\|\,\tau(A) = \tau(A) .$$

Lemma 9. *For any pair of vectors* $\varphi$ and $\psi$ *we have,*

(i) $\varphi \otimes \overline{\psi}$ *is in* $(\tau c)$.
(ii) $t(\varphi \otimes \overline{\psi}) = (\varphi, \psi)$.
(iii) $\tau(\varphi \otimes \overline{\psi}) = \|\varphi\| \, \|\psi\|$ .

*Proof.* We assume $\varphi \neq 0$ and $\psi \neq 0$ otherwise the proof is trivial. Clearly, $\varphi_1 = \frac{\psi}{\|\psi\|}$ has norm 1 and is a proper vector of

$$[\varphi \otimes \overline{\psi}]^2 = (\varphi \otimes \overline{\psi})^* (\varphi \otimes \overline{\psi}) = \|\varphi\|^2 \psi \otimes \overline{\psi}$$

corresponding to the proper value $\|\varphi\|^2 \|\psi\|^2$. Furthermore, every vector orthogonal to $\varphi_1$, that is to $\psi$, is also a proper vector of $[\varphi \otimes \overline{\psi}]^2$ corresponding to the proper value 0. We choose a basis $\{\varphi_j\}$ in our space having $\varphi_1$ as one of its elements. Then,

$$[\varphi \otimes \overline{\psi}]\varphi_1 = \|\varphi\| \, \|\psi\| \, \varphi_1$$

and

$$[\varphi \otimes \overline{\psi}]\varphi_j = 0 \qquad \text{for } j \neq 1.$$

Consequently,

$$\Sigma_j \left([\varphi \otimes \overline{\psi}]\varphi_j, \varphi_j\right) = \|\varphi\| \, \|\psi\| < +\infty .$$

This means that $[\varphi \otimes \overline{\psi}]$ is in $(\tau c)$ and (iii) holds. Consequently $\varphi \otimes \overline{\psi}$ is also in $(\tau c)$. Moreover, Parseval's identity gives

$$t(\varphi \otimes \overline{\psi}) = \Sigma_j \left((\varphi \otimes \overline{\psi})\varphi_j, \varphi_j\right) = \Sigma_j (\varphi_j, \psi)(\varphi, \varphi_j) = (\varphi, \psi) .$$

Theorem 1. *Every operator in the trace-class is necessarily in the Schmidt-class, hence completely continuous.*

*Proof.* This is a consequence of Definition 1 and Theorem II.1.

Theorem 2. *Let* $A$ *be a completely continuous operator and* $\Sigma_i \lambda_i \varphi_i \otimes \overline{\psi}_i$ *its polar form. Then* $A \in (\sigma c)$ *if and only if* $\Sigma_i \lambda_i^2 < +\infty$; *in the last case* $\sigma(A) = (\Sigma_i \lambda_i^2)^{\frac{1}{2}}$. *Furthermore,* $A \in (\tau c)$ *if and only if* $\Sigma_i \lambda_i < +\infty$; *in the last case* $\tau(A) = \Sigma_i \lambda_i$.

*Proof.* The first statement forms the content of Theorem II.2, and is repeated here merely for the sake of comparison. Now $A \in (\tau c)$ is equivalent to $[A] = \Sigma_i \lambda_i \psi_i \otimes \overline{\psi}_i$ is in $(\tau c)$. The last — since $[A]\,\psi = 0$ for every $\psi$ orthogonal to the sequence $\{\psi_i\}$ — is equivalent to

$$\tau(A) = \Sigma_i \left([A]\psi_i, \psi_i\right) = \Sigma_i \lambda_i < +\infty .$$

Theorem 3. *Let* $A \in (\tau c)$. *For a given* $\varepsilon > 0$, *we can find an operator* $B$ *of finite rank such that* $\tau(A - B) < \varepsilon$.

*Proof.* Let $A = \Sigma_i \lambda_i \varphi_i \otimes \overline{\psi}_i$. By the previous theorem $\Sigma_i \lambda_i < +\infty$. Put $A_n = \Sigma_{i=1}^n \lambda_i \varphi_i \otimes \overline{\psi}_i$. Then

$$\tau(A - A_n) = \Sigma_{i>n} \lambda_i .$$

For large $n$, the right side is $< \varepsilon$.

Theorem 4. *We have*

$$\|A\| \leqq \sigma(A) \leqq \tau(A) .$$

*Proof.* In view of Lemma II.2, it is sufficient to prove that $\sigma(A) \leqq \tau(A)$. We may only consider the case $\tau(A) < +\infty$, that is, $A \in (\tau c)$. Writing $A = \Sigma_i \lambda_i \varphi_i \otimes \overline{\psi}_i$ we have,

$$\sigma(A) = (\Sigma_i \lambda_i^2)^{\frac{1}{2}} \leqq \Sigma_i \lambda_i = \tau(A) .$$

We summarize a part of the preceding discussion:

Theorem 5. *Consider the class* ($\tau c$) *of all products of two operators in* ($\sigma c$). *This class coincides with the class of all those operators* $A$ *for which* $\Sigma_j([A]\varphi_j, \varphi_j) < +\infty$ *for a fixed basis* $\{\varphi_j\}$. *With the obvious definition of addition and scalar multiplication* ($\tau c$) *is a linear space. The last is normed if the above sum — which is independent on the chosen basis* $\{\varphi_j\}$ *— stands for the norm* $\tau(A)$ *of an operator* $A$. *The resulting normed linear space is complete; it contains the operators of finite rank as a dense subset. The operators in* ($\tau c$) *necessarily belong to* ($\sigma c$), *hence are completely continuous; they are precisely the operators* $A$ *having the polar form* $\Sigma_i \lambda_i \omega_i \otimes \overline{\psi}_i$ *with* $\Sigma_i \lambda_i < +\infty$; *we then have* $\tau(A) = \Sigma_i \lambda_i$. *Moreover* ($\tau c$) *is a (two-sided) ideal in the algebra* $\mathfrak{A}$, *and a Banach algebra under its own norm. In fact,* ($\tau c$) *is also a norm ideal.*

Remark. To avoid repetition of an argument, we do not bother at present to prove that ($\tau c$) is complete. This will follow from a more general theorem proven later. We shall also prove that ($\tau c$) may be interpreted as the conjugate space of the Banach space $\mathfrak{C}$ of all completely continuous operators and thus is necessarily complete.

Remark. For an operator $A \in (\tau c)$ we have $\tau(A) = \Sigma_j([A]\varphi_j, \varphi_j)$, where $\{\varphi_j\}$ stands for an arbitrary basis. The sum on the right not only involves the inner product but also $[A]$ — notions which are meaningful only in Hilbert spaces. It is interesting that for operators $A$ of finite rank, $\tau(A)$ may be expressed also in a language which is meaningful in perfectly general Banach spaces, permitting hereby to carry over to those spaces the concept of the trace-class; the last is defined as the metric completion of the linear manifold of all operators of finite rank on which $\tau(A)$ represents the norm. This definition is justified by Theorem 3.

Consider an operator $A$ of finite rank and let $\Sigma_{j=1}^n \lambda_j \varphi_j \otimes \overline{\psi}_j$ be its polar representation. Clearly, $A$ admits also many other representations $\Sigma_{i=1}^m f_i \otimes \bar{g}_i$ with families $f_1, \ldots, f_m$ and $g_1, \ldots, g_m$ not necessarily orthonormal. For each such representation of $A$ we have

$$\Sigma_{j=1}^n \lambda_j = \tau(A) = \tau(\Sigma_{i=1}^m f_i \otimes \bar{g}_i) \leqq \Sigma_{i=1}^m \tau(f_i \otimes \bar{g}_i) = \Sigma_{i=1}^m \|f_i\| \, \|g_i\| .$$

This proves that

$$\tau(A) = \inf \Sigma_{i=1}^m \|f_i\| \, \|g_i\|$$

where the infimum extends over the set of all sums corresponding to all representations of the operator $A$ of finite rank. The last equation defines $\tau(A)$ for all $A \in \mathfrak{R}$ in a language which carries over to operators of finite rank from one Banach space into another.

## 2. A maximum problem in $(\tau c)$

Lemma 10. *Let $A \in (\tau c)$. Then $t(XA) = 0$ for all Hermitean $X$ if and only if $A = 0$.*

*Proof.* Since any $X$ may be written in the form $X = X_1 + iX_2$ with both $X_1$ and $X_2$ Hermitean, our condition is equivalent to

$$t(XA) = 0$$

for all $X$.

Whenever it is satisfied we have in particular, $(A\varphi, \psi) = t((\varphi \otimes \overline{\psi})A) = 0$ for all pairs of vectors $\varphi$ and $\psi$ and therefore $A = 0$.

Corollary. *Let $A \in (\tau c)$. Then $t(XA)$ is real for all Hermitean $X$ if and only if $A$ itself is Hermitean.*

*Proof.* Since

$$\overline{t(XA)} = t(A^*X^*) = t(X^*A^*)\,,$$

our condition is equivalent to

$$t(X(A - A^*)) = 0 \quad \text{for all Hermitean } X.$$

This will happen if and only if $A - A^* = 0$.

Theorem 6. *Let $A \in (\tau c)$. Then*

$$\mathcal{R}\, t(A) \geqq \mathcal{R}\, t(UA)$$

*for all unitary $U$ if and only if $A \geqq 0$.*

*Proof.* Assume $A \geqq 0$. Then $A = [A]$ and thus $t(A) = t([A]) = \tau(A)$. We have,

$$\mathcal{R}\, t(UA) \leqq |t(UA)| \leqq \tau(UA) \leqq \|U\|\ \tau(A) = \tau(A) = t(A) = \mathcal{R} t(A)\,.$$

Conversely. Assume that $\mathcal{R} t(A) \geqq \mathcal{R} t(UA)$ for all unitary $U$.

Let $X$ be Hermitean. Clearly, if $\varepsilon \gtrless 0$ is small enough $I \pm i\varepsilon X$ is invertible. Denoting for brevity $i\varepsilon X$ by $Y$, one has $Y^* = -Y$ and

$$(I - Y)^{-1} = I + Y + Y^2 + \cdots$$
$$(I + Y)^{-1} = I - Y + Y^2 - \cdots;$$

the two series on the right converge (uniformly) in bound. Defining $U$ and $V$ by means of the relations

$$U = (I + Y)\,(I - Y)^{-1} = I + 2\,Y + 2\,Y^2 + \cdots$$
$$V = (I - Y)\,(I + Y)^{-1} = I - 2\,Y + 2\,Y^2 - \cdots$$

one has $V = U^*$ and

$$U U^* = U V = I$$

$$U^* U = V U = I .$$

Thus, if $X$ is Hermitean, then for small $\varepsilon \gtrless 0$

$$U = (I + i \varepsilon X) (I - i \varepsilon X)^{-1}$$

is unitary. Moreover,

$$U = I + 2 i \varepsilon X + \varepsilon^2 T(\varepsilon)$$

where

$$T(\varepsilon) = 2 (i X)^2 (I + Y + Y^2 + \cdots)$$

is a well defined operator depending on $\varepsilon$ such however that

$$\|T(\varepsilon)\| \leqq 4 \|X\|^2 \quad \text{whenever} \quad |\varepsilon| < \frac{1}{2\|X\|} .$$

Since in particular $\mathscr{R}t(A) \geqq \mathscr{R}t(UA)$ also holds for the above $U$ one gets

$$\mathscr{R}t(A) \geqq \mathscr{R}t(A) + 2 \varepsilon \mathscr{R}t(iXA) + \varepsilon^2 \mathscr{R}t(T(\varepsilon) A) .$$

Thus, for $\varepsilon \gtrless 0$ sufficiently small

$$2 \varepsilon \mathscr{R}t(iXA) + \varepsilon^2 \mathscr{R}t(T(\varepsilon) A) \leqq 0 .$$

Since

$$|\mathscr{R}t(T(\varepsilon) A)| \leqq \|T(\varepsilon)\| \, \tau(A) \leqq 4 \|X\|^2 \tau(A)$$

it necessarily follows

$$\mathscr{R}t(iXA) = 0 .$$

Thus, $t(XA)$ is real for all Hermitean $X$. By Lemma 10 (Corollary), $A$ is Hermitean.

Being Hermitean and in ($\tau c$), $A$ may be written in the spectral form $\Sigma_i \lambda_i \varphi_i \otimes \overline{\varphi}_i$ where all the $\lambda_i$'s are real, $\neq 0$ and $\Sigma_i |\lambda_i| < +\infty$. The inequality $\mathscr{R}t(A) \geqq \mathscr{R}t(UA)$ therefore amounts to

$$\Sigma_i \lambda_i \geqq \mathscr{R} \, \Sigma_i \lambda_i (U\varphi_i, \varphi_i)$$

for all unitary $U$. Choosing in particular $U$ so that $U\varphi_i = \varphi_i$ if $\lambda_i > 0$ and $U\varphi_i = -\varphi_i$ if $\lambda_i < 0$, one has

$$\Sigma_i \lambda_i \geqq \Sigma_i |\lambda_i| .$$

Consequently, all the $\lambda_i$ are $> 0$ and thus $A \geqq 0$.

Corollary. *Let $A \in (\tau c)$. Then $U_0$ is maximal for $\mathscr{R}t(UA)$ if and only if $U_0 A = [A]$.*

*Proof.* $\mathscr{R}t(U_0 A) \geqq \mathscr{R}t(UA)$ for all $U$ is equivalent to (replacing $U$ by $U U_0$) $\mathscr{R}t(U_0 A) \geqq \mathscr{R}t(U U_0 A)$ for all $U$. By the previous theorem the last will happen if and only if $U_0 A \geqq 0$. Now, $U_0 A \geqq 0$ implies

$$(U_0 A)^2 = (U_0 A)^* (U_0 A) = A^* U_0^* U_0 A = A^* A .$$

Thus, $U_0 A = [A]$.

Corollary. *Let $A \in (\tau c)$. A unitary operator $U_0$ satisfies the equality*

$$\mathscr{R}t(U_0 A) = \tau(A)$$

*if and only if $U_0 A = [A]$.*

*Proof.* Clearly, $|\mathscr{R}t(UA)| \leqq \tau(A)$ for all unitary $U$. The above equality implies thus that $U_0$ is maximal for $\mathscr{R}t(UA)$. Consequently $U_0 A = [A]$. The converse is obvious.

Remark. We supplement the last discussion with an example of an operator $A \in (\tau c)$ for which there is no unitary $U$ which maximizes $\mathscr{R}t(UA)$ or equivalently, which satisfies the equality $UA = [A]$. For this purpose choose a separable space. If $\{\varphi_i\}$ is a complete orthonormal sequence and $\{\lambda_i\}$ is a sequence of positive numbers with $\Sigma_i \lambda_i < +\infty$, then the operator

$$A = \Sigma_i \lambda_i \, \varphi_{i+1} \otimes \overline{\varphi}_i$$

satisfies our requirements. The equality $WA = [A]$ amounts to

$$W(\Sigma_i \lambda_i \varphi_{i+1} \otimes \overline{\varphi}_i) = \Sigma_i \lambda_i \varphi_i \otimes \overline{\varphi}_i \, .$$

Operating both sides on $\varphi_i$, one gets $\lambda_i W \varphi_{i+1} = \lambda_i \varphi_i$, that is,

$$W \varphi_{i+1} = \varphi_i \qquad \text{for } i = 1, 2, \ldots$$

Thus, $W$ must not be unitary.

Corollary. *Let $A$ or $B$ belong to $(\tau c)$. Whenever,*

$$\mathscr{R}t(AB) \geqq \mathscr{R}t(UAVB)$$

*for all unitary pairs $U$, $V$, then both $AB$ and $BA$ are $\geqq 0$.*

Proof. Clearly, both $AB$ and $BA$ belong to $(\tau c)$. Replacing $V$ by the identity, our assumption implies that $\mathscr{R}t(AB) \geqq \mathscr{R}t(UAB)$ for all unitary $U$. Similarly, replacing $U$ by the identity, one gets $\mathscr{R}t(BA) = \mathscr{R}t(AB) \geqq \mathscr{R}t(AVB) = \mathscr{R}t(VBA)$ for all unitary $V$. An application of Theorem 6 concludes the proof.

# IV. The successive conjugate spaces of the space $\mathfrak{C}$ of all completely continuous operators

## 1. A characterization of $\mathfrak{C}^*$ and $\mathfrak{C}^{**}$

We denote by $\mathfrak{B}$ the Banach space (algebra) of all operators on $\mathfrak{H}$ and by $\mathfrak{C}$ its subspace (Banach subalgebra) of all completely continuous operators; the bound of an operator represents its norm.

Lemma 1. *Let $A$ and $B$ be two operators. Then*

$$t(XA) = t(XB)$$

*for all operators $X$ of rank 1 if and only if $A = B$.*

*Proof.* This is true since

$$0 = t((\varphi \otimes \overline{\psi})A) = (A\varphi, \psi)$$

for all pairs of vectors $\varphi$ and $\psi$ if and only if $A = 0$.

Theorem 1. *Let $T$ be a fixed operator in $(\tau c)$. As $C$ varies in $\mathfrak{C}$ the expression*

$$t(C\,T)$$

*represents a bounded linear functional on $\mathfrak{C}$; its bound is equal to $\tau(T)$. Moreover, every bounded linear functional on $\mathfrak{C}$ is obtained in such a manner (from a unique $T$ in $(\tau c)$).*

*This establishes a one-to-one correspondence preserving the linearity relations and norm between the spaces $\mathfrak{C}^*$ and $(\tau c)$. Hence, $\mathfrak{C}^*$ and $(\tau c)$ are equivalent (i. e., modulo the language of Banach identical).*

*Proof.* Let $\mathscr{F}$ be a linear bounded functional on $\mathfrak{C}$, and $\|\mathscr{F}\|$ stand for its bound. Consider $\mathscr{F}(\varphi \otimes \overline{\psi})$ for all pairs of vectors $\varphi$ and $\psi$. Clearly,

(i) $\mathscr{F}(\varphi \otimes \overline{(c\,\psi)}) = \bar{c}\mathscr{F}(\varphi \otimes \overline{\psi})$.
(i') $\mathscr{F}((c\,\varphi) \otimes \overline{\psi}) = c\mathscr{F}(\varphi \otimes \overline{\psi})$.
(ii) $\mathscr{F}(\varphi \otimes \overline{(\psi_1 + \psi_2)}) = \mathscr{F}(\varphi \otimes \overline{\psi}_1) + \mathscr{F}(\varphi \otimes \overline{\psi}_2)$.
(ii') $\mathscr{F}((\varphi_1 + \varphi_2) \otimes \overline{\psi}) = \mathscr{F}(\varphi_1 \otimes \overline{\psi}) + \mathscr{F}(\varphi_2 \otimes \overline{\psi})$.
(iii) $|\mathscr{F}(\varphi \otimes \overline{\psi})| \leqq \|\mathscr{F}\|\,\|\varphi \otimes \overline{\psi}\| = \|\mathscr{F}\|\,\|\varphi\|\,\|\psi\|$.

Holding $\varphi$ fixed and varying $\psi$ — (i), (ii) and (iii) show that $\mathscr{F}(\varphi \otimes \overline{\psi})$ determines a complex conjugate of a bounded linear functional on $\mathfrak{H}$. Hence by the well-known lemma of F. RIESZ on the representation of bounded linear functionals, there exists a unique element $\varphi'$ such that $\mathscr{F}(\varphi \otimes \overline{\psi}) = (\varphi', \psi)$. Define $T\varphi = \varphi'$. Thus,

$$\mathscr{F}(\varphi \otimes \overline{\psi}) = (T\varphi, \psi)$$

for all pairs $\varphi$, $\psi$ in $\mathfrak{H}$. Because of (i') and (ii'), $T$ is linear. It is also bounded hence an operator since,

$$\|T\| = \sup_{\|\varphi\| = \|\psi\| = 1} |(T\varphi, \psi)| = \sup_{\|\varphi\| = \|\psi\| = 1} |\mathscr{F}(\varphi \otimes \overline{\psi})| \leqq \|\mathscr{F}\|\,.$$

If $X$ is of finite rank say $X = \Sigma_{i=1}^{n} f_1 \otimes \bar{g}_i$ (the $f_1, \ldots, f_n$ and $g_1, \ldots, g_n$ are not necessarily orthonormal) we have

$$\mathscr{F}(X) = \Sigma_{i=1}^{n} \mathscr{F}(f_i \otimes \bar{g}_i) = \Sigma_{i=1}^{n} (Tf_i, g_i)$$
$$= \Sigma_{i=1}^{n} t(Tf_i \otimes \bar{g}_i) = t(TX) = t(XT)\,.$$

The above defined $T$ is not only an operator but also belongs to $(\tau c)$. To prove this, we use the polar decomposition $T = W[T]$, $[T] = W^*T$ and choose a basis $\{\varphi_j\}$ in the initial set of $W$. For any finite set $J$ of indices $j$,

$$\|\Sigma_{j \in J}\, \varphi_j \otimes \overline{W\varphi_j}\| = 1$$

and therefore,

$$\Sigma_{j \in J}([T]\varphi_j, \varphi_j) = \Sigma_{j \in J}(W^* T\varphi_j, \varphi_j)$$
$$= \Sigma_{j \in J}(T\varphi_j, W\varphi_j) = t((\Sigma_{j \in J} \varphi_j \otimes \overline{W\varphi_j})T)$$
$$= \mathscr{F}(\Sigma_{j \in J} \varphi_j \otimes \overline{W\varphi_j}) \leqq \|\mathscr{F}\| .$$

Moreover, if $\varphi$ is orthogonal to all the $\varphi_j$ we have $W\varphi = 0$ and therefore $([T]\varphi, \varphi) = (T\varphi, W\varphi) = 0$. Consequently, $T \in (\tau c)$ and

$$\tau(T) \leqq \|\mathscr{F}\| .$$

Thus $T \in (\tau c)$ and therefore $t(AT)$ is defined for all operators $A$ and in particular for every completely continuous operator. Now given a $C \in \mathfrak{C}$, choose a sequence $\{X_n\}$ of operators of finite rank such that $\|C - X_n\| \to 0$. Then $\mathscr{F}(X_n) \to \mathscr{F}(C)$. On the other hand

$$|t(CT) - t(X_n T)| \leqq \tau(T)\, \|C - X_n\|$$

implies $t(X_n T) \to t(CT)$. Thus,

$$\mathscr{F}(C) = t(CT) \qquad \text{for all } C \in \mathfrak{C}$$

and therefore also

$$\|\mathscr{F}\| \leqq \tau(T)$$

since,

$$|t(CT)| \leqq \tau(T)\, \|C\| .$$

Conversely. Let $T$ be a fixed operator in $(\tau c)$. Then $t(CT)$ is obviously a linear functional on $\mathfrak{C}$. The last inequality shows that this functional is bounded on $\mathfrak{C}$. By what was already shown, its bound must be equal to $\tau(T)$.

Corollary. $(\tau c)$ *is complete.*

*Proof.* This is true since $(\tau c)$ is the conjugate space of the Banach space $\mathfrak{C}$.

Theorem 2. *Let $A$ be a fixed operator. As $T$ varies in $(\tau c)$, the expression*

$$t(TA)$$

*represents a bounded linear functional on $(\tau c)$; its bound is given by $\|A\|$. Moreover, every bounded linear functional on $(\tau c)$ is obtained in such a manner. This establishes a one-to-one correspondence preserving linearity and the norm between the Banach spaces $(\tau c)^*$ and $\mathfrak{B}$. Hence $(\tau c)^*$ and $\mathfrak{B}$ are equivalent (that is, in the language of Banach identical).*

*Proof.* Let $A$ be fixed. Then $t(TA)$ is defined for all $T \in (\tau c)$ and is obviously linear. The inequality $|t(TA)| \leqq \|A\|\, \tau(T)$ implies that this functional has on $(\tau c)$ a bound $\leqq \|A\|$. On the other hand, for any two vectors $\varphi, \psi$, both of norm 1, we have $\tau(\varphi \otimes \overline{\psi}) = 1$ and $|t((\varphi \otimes \overline{\psi})A)| = |(A\varphi, \psi)|$. The bound of the above functional is thus precisely $\|A\|$.

Conversely. Let $\mathscr{F}$ be a bounded linear functional on $(\tau c)$. As in the proof of the previous theorem, we use F. RIESZ' representation theorem for bounded linear functionals and determine the unique operator $A$ such that $\mathscr{F}(X) = t(XA)$ for all $X$ of finite rank. Since the operators of finite rank form a dense set in $(\tau c)$, the two bounded linear functionals $\mathscr{F}(T)$ and $t(TA)$ must coincide for all $T \in (\tau c)$. By what was already proved the bound of $\mathscr{F}$ is equal to $\|A\|$.

Finally, we remark that the established one-to-one correspondence between the operators and the bounded linear functionals on $(\tau c)$, preserves addition and scalar multiplication. This concludes the proof.

We summarize the preceding discussion:

Theorem 3. *The linear set of all completely continuous operators on $\mathfrak{H}$ where the bound of an operator is considered as its norm, furnishes a Banach space $\mathfrak{C}$. Its first conjugate space $\mathfrak{C}^*$ may be interpreted as the trace-class. The trace-class is the Banach space of all operators $A$ on $\mathfrak{H}$, for which $\Sigma_j([A]\varphi_j, \varphi_j) < +\infty$ for a complete orthonormal family $\{\varphi_j\}$; the last sum is independent on the chosen basis and represents the norm of $A$ in $\mathfrak{C}^*$. The second conjugate space $\mathfrak{C}^{**}$ of $\mathfrak{C}$ may be interpreted as the Banach space of all operators on $\mathfrak{H}$ where again the bound of an operator stands for its norm.*

Corollary. *If $\mathfrak{H}$ is infinite-dimensional, then $\mathfrak{C}$ is non-reflexive.*

*Proof.* Consider the "natural" imbedding of $\mathfrak{C}$ into $\mathfrak{C}^{**}$. A fixed element $C_0 \in \mathfrak{C}$ determines the linear bounded functional

$$t(C_0T) = t(TC_0)$$

on $(\tau c) = \mathfrak{C}^*$. From the above discussion it is clear that this functional again corresponds to $C_0$ in our identification of $\mathfrak{C}^{**} = (\tau c)^*$ with $\mathfrak{B}$.

## 2. $\mathfrak{C}$ is not a conjugate space

We assume throughout this section that $\mathfrak{H}$ is *infinite-dimensional.*

Put $\mathfrak{C}^{(1)} = \mathfrak{C}^*$ and $\mathfrak{C}^{(n)} = (\mathfrak{C}^{(n-1)})^*$ for $n > 1$. By its very construction a Banach space is a closed subset of any metric space in which it may be imbedded. Thus, $\mathfrak{C}$ is a proper subspace not only of $\mathfrak{C}^{(2)}$ but also of $\mathfrak{C}^{(2n)}$ for $n = 1, 2, 3, \ldots$ Since a subspace of a reflexive Banach space is necessarily also reflexive (see PETTIS [1]), it follows that none of the spaces $\mathfrak{C}^{(2n)}$ is reflexive, that is,

$$\mathfrak{C} \subset \mathfrak{C}^{(2)} \subset \mathfrak{C}^{(4)} \subset \ldots$$

where each inclusion is a proper one. Moreover, since a Banach space is reflexive if and only if its conjugate space is reflexive (see PETTIS [1]), none of the spaces $\mathfrak{C}^{(2n-1)}$ $n = 1, 2, 3, \ldots$ is reflexive, that is,

$$(\tau c) = \mathfrak{C}^{(1)} \subset \mathfrak{C}^{(3)} \subset \mathfrak{C}^{(5)} \subset \ldots$$

where again each inclusion is a proper one.

It is interesting to note that the space $\mathfrak{C}$ actually forms a "beginning" in the above chain of inclusions. The argument follows.

Let $\mathscr{B}$ be a Banach space. A set $\mathscr{K} \subset \mathscr{B}^*$ is termed regularly convex if for every $f_0^* \in \mathscr{B}^*$ and not in $\mathscr{K}$ there is an $f_0 \in \mathscr{B}$ such that

$$\sup_{f^* \in \mathscr{K}} f^*(f_0) < f_0^*(f_0) \, .$$

It is clear that every regularly convex subset of $\mathscr{B}^*$ is necessarily convex. The solid unit sphere of $\mathscr{B}^*$ is regularly convex.

A theorem due to KREIN-MILMAN [1] states: *If $\mathscr{K}$ is a bounded, regularly convex subset of $\mathscr{B}^*$, then the set $\mathfrak{S}$ of its extreme points is non-empty and $\mathscr{K}$ coincides with the smallest regularly convex set containing $\mathfrak{S}$.* Thus, if the solid unit sphere of a Banach space $\mathscr{B}$ has no extreme points, then $\mathscr{B}$ must not be the conjugate space (of any Banach space). In particular, if $\mathscr{B}$ is such that for any $f \in \mathscr{B}$ with $\|f\| \leqq 1$ there is a non-zero $g \in \mathscr{B}$ such that $\|f \pm g\| \leqq 1$ then $\mathscr{B}$ must not be a conjugate space. The relation

$$f = \frac{(f+g)+(f-g)}{2}$$

then proves that $f$ is the midpoint of the segment formed by the vectors $f+g$ and $f-g$. Consequently, $f$ is not an extreme point of the solid unit sphere of $\mathscr{B}$.

Lemma 2. *If $\mathfrak{H}$ is infinite-dimensional, then the solid unit sphere of $\mathfrak{C}$ has no extreme points: For a given completely continuous operator $A$ with a bound $\|A\| \leqq 1$ we can always find an operator $B$ of rank 1 (hence completely continuous) with $\|B\| \leqq 1$, for which also $\|A+B\| \leqq 1$ and $\|A-B\| \leqq 1$.*

*Proof.* It will be convenient to consider three separate cases.

1. $A = 0$. In this case we may choose for $B$ any operator of rank 1 with a bound $\leqq 1$.

2. $A$ is of finite rank, $A = \Sigma_{i=1}^{n} \lambda_i \varphi_i \otimes \overline{\psi}_i$. Then,

$$\|A\| \leqq 1 \quad \text{if and only if} \quad \max\{\lambda_1, \ldots, \lambda_n\} \leqq 1 \, .$$

Choose two normal vectors $\varphi$ and $\psi$ orthogonal to $\varphi_1, \ldots, \varphi_n$ and $\psi_1, \ldots, \psi_n$ respectively and set $B = \varphi \otimes \overline{\psi}$. Then, $\|B\| = 1$ and

$$\|A \pm B\| = \max\{\lambda_1, \ldots, \lambda_n, 1\} = 1 \, .$$

3. The range of $A$ is infinite-dimensional. Let $A = \Sigma_i \lambda_i \varphi_i \otimes \overline{\psi}_i$; we have $0 < \lambda_i \leqq 1$ and $\lambda_i \to 0$. Choose a fixed $i_0$ for which $\lambda_{i_0} < 1$ and a positive $\varepsilon$ for which $\lambda_{i_0} + \varepsilon < 1$. The operator $B = \varepsilon \varphi_{i_0} \otimes \overline{\psi}_{i_0}$ will satisfy our requirement since $\|B\| = \varepsilon$ and

$$\|A \pm B\| = \sup\{\lambda_1, \lambda_2, \ldots, \lambda_{i_0 - 1}, |\lambda_{i_0} \pm \varepsilon|, \lambda_{i_0+1}, \ldots\} \leqq 1 \, .$$

This concludes the proof.

Theorem 4. *If $\mathfrak{H}$ is infinite-dimensional, then $\mathfrak{C}$ is not the conjugate space of any Banach space.*

*Proof.* The proof is a consequence of Lemma 2 and the Krein-Milman theorem referred to above.

## 3. Some linear functionals on the space of all operators

We assume throughout this section that $\mathfrak{H}$ is *infinite-dimensional.*

Below we state some results concerning $\mathfrak{C}^{(3)} = \mathfrak{B}^*$. In particular we characterize its subspace which may be identified with $(\tau c)$. There is little however that we can say about the spaces $\mathfrak{C}^{(n)}$ for $n \geqq 4$. Of course, each of those spaces represents "the space of all linear bounded functionals on the space of all linear bounded functionals . . .". This kind of information — a multiple restatement of the definition of a conjugate space — leaves us only bewildered. What one should be after, is a direct simple characterization of $\mathfrak{C}^{(n)}$ (not involving the help of the $\mathfrak{C}^{(k)}$ for $k < n$) as in the case of $\mathfrak{C}^{(1)} = (\tau c)$ or $\mathfrak{C}^{(2)} = \mathfrak{B}$.

If $T \in (\tau c)$, the expression $t(TB)$ determines a bounded linear functional on $\mathfrak{B}$ or $\mathfrak{C}$ depending whether we permit $B$ to vary in $\mathfrak{B}$ or in $\mathfrak{C}$. In both cases the bound is the same, namely $\tau(T)$. We also remark that $0 = t(T(\varphi \otimes \overline{\psi})) = (T\varphi, \psi)$ for all pairs of vectors $\varphi$ and $\psi$ implies $T = 0$ and hence $t(TB) = 0$ for all $B \in \mathfrak{B}$.

Theorem 1 states that all bounded linear functionals on $\mathfrak{C}$ are of the form $t(TB)$ and that $\mathfrak{C}^*$ may be identified with $(\tau c)$. On the other hand if $\mathfrak{H}$ is infinite-dimensional, only some bounded linear functionals on $\mathfrak{B}$ are of the special form $\mathscr{F}(B) = t(TB)$ with $T \in (\tau c)$ (precisely those which are in the subspace of $\mathfrak{B}^*$ determined by the natural imbedding of $(\tau c)$ in its second conjugate space). There is no possibility of misunderstanding if we conveniently write $\mathscr{F} \in (\tau c)$ for every bounded linear functional $\mathscr{F}$ on $\mathfrak{B}$ of the above form. We also denote by $\mathfrak{C}^{\perp}$ the subspace of $\mathfrak{B}^*$ consisting of all the bounded linear functionals on $\mathfrak{B}$ which vanish identically on $\mathfrak{C}$. By the remark stated in the preceding paragraph $(\tau c)$ and $\mathfrak{C}^{\perp}$ have only the 0-functional in common.

Let $\mathscr{F}$ be a bounded linear functional on $\mathfrak{B}$. Its restriction to $\mathfrak{C}$ is by Theorem 1 of the form $t(TB)$ where $T$ is fixed in $(\tau c)$ and $B$ varies in $\mathfrak{C}$. Permitting $B$ to vary in $\mathfrak{B}$, the same expression $t(TB)$ determines an element $\mathscr{F}_1 \in (\tau c)$. Of course $\mathscr{F} - \mathscr{F}_1$ is in $\mathfrak{C}^{\perp}$.

Theorem 5, (Dixmier [1]). *Every bounded linear functional $\mathscr{F}$ on $\mathfrak{B}$ may be represented in one and only one way in the form $\mathscr{F} = \mathscr{F}_1 + \mathscr{F}_2$ where $\mathscr{F}_1 \in (\tau c)$ and $\mathscr{F}_2 \in \mathfrak{C}^{\perp}$. Moreover, $\|\mathscr{F}\| = \|\mathscr{F}_1\| + \|\mathscr{F}_2\|$.*

*Proof.* The existence of the decomposition was proven above. It is unique since $(\tau c)$ and $\mathfrak{C}^{\perp}$ have only the 0-functional in common. Since

$\|\mathscr{F}\| \leqq \|\mathscr{F}_1\| + \|\mathscr{F}_2\|$ it is sufficient to show $\|\mathscr{F}\| \geqq \|\mathscr{F}_1\| + \|\mathscr{F}_2\| - \varepsilon$ for any $\varepsilon > 0$.

Clearly $\mathscr{F}_1$ on $\mathfrak{B}$ and its restriction to $\mathfrak{C}$ have the same bound $\|\mathscr{F}_1\|$. Suppose that $\mathscr{F}_1$ is determined by the operator $T = \Sigma_i \lambda_i \varphi_i \otimes \overline{\psi}_i$ with $\Sigma_i \lambda_i < +\infty$, that is,

$$\mathscr{F}_1(B) = t(TB) = \Sigma_i \lambda_i (B\varphi_i, \psi_i) .$$

Given $\varepsilon > 0$, choose

1) $n$ so large that $\Sigma_{i>n} \lambda_i < \frac{\varepsilon}{3}$
2) $A^0$ of finite rank with $\|A^0\| = 1$ and $\mathscr{F}_1(A^0) > \|\mathscr{F}_1\| - \frac{\varepsilon}{3}$
3) $\widetilde{B} \in \mathfrak{B}$ with $\|\widetilde{B}\| = 1$ and $\mathscr{F}_2(\widetilde{B}) > \|\mathscr{F}_2\| - \frac{\varepsilon}{3}$.

Let $P$ be the projection on the finite-dimensional subspace spanned by $\varphi_1, \ldots, \varphi_n; \psi_1, \ldots, \psi_n$ and the ranges of both $A^0$ and $(A^0)^*$. Define $B^0 = (I - P)\widetilde{B}(I - P)$. We obviously have $\|B^0\| \leqq \|\widetilde{B}\| = 1$ and $B^0 \varphi_i = B^0 \psi_i = 0$ for $i = 1, \ldots, n$. Consequently,

$$|\mathscr{F}_1(B^0)| = |\Sigma_{i>n} \lambda_i (B^0\varphi_i, \psi_i)| \leqq \Sigma_{i>n} \lambda_i < \frac{\varepsilon}{3} .$$

Clearly, $\widetilde{B} - B^0 = P\widetilde{B} + \widetilde{B}P - P\widetilde{B}P$ is of finite rank and thus $\mathscr{F}_2(\widetilde{B}) = \mathscr{F}_2(B^0)$. Since also $\mathscr{F}_2(A^0) = 0$, we have

$$|\mathscr{F}(A^0 + B^0)| = |\mathscr{F}_1(A^0) + \mathscr{F}_1(B^0) + \mathscr{F}_2(B^0)| \geqq$$
$$\geqq \mathscr{F}_1(A^0) + \mathscr{F}_2(B^0) - |\mathscr{F}_1(B^0)| > \|\mathscr{F}_1\| + \|\mathscr{F}_2\| - \varepsilon .$$

Moreover, since

$$\|A^0 + B^0\| = \sup\{\|A^0\|, \|B^0\|\} = 1$$

we have

$$\|\mathscr{F}\| \geqq \|\mathscr{F}_1\| + \|\mathscr{F}_2\| - \varepsilon .$$

Let $\mathscr{L}$ be a linear space and $\mathscr{L}^+$ stand for the space of all linear functionals on $\mathscr{L}$. A linear manifold $\mathfrak{S}$ of $\mathscr{L}^+$ is termed *total* if $f \in \mathscr{L}$ and $\mathscr{F}(f) = 0$ for all $\mathscr{F} \in \mathfrak{S}$ implies $f = 0$.

Let $\mathfrak{S} \subset \mathscr{L}^+$ be total. By the *$\mathfrak{S}$-topology* in $\mathscr{L}$ we understand the weakest topology in which all the functionals in $\mathfrak{S}$ are continuous. Obviously, this is the topology for which a subbase is given by all sets of the form

$$\{f \in \mathscr{L} : |\mathscr{F}(f) - \mathscr{F}(f_0)| < \varepsilon\}$$

where $\mathscr{F} \in \mathfrak{S}$, $f_0 \in \mathscr{L}$ and $\varepsilon > 0$; this means that the open sets in $\mathscr{L}$ are precisely the unions of finite intersections of sets of the above type. $\mathscr{L}$ is then a linear topological (Hausdorff) space which is locally convex (that is, it has a basis for its topology consisting of convex sets). Moreover, the linear functionals on $\mathscr{L}$ which are continuous in the $\mathfrak{S}$-topology are precisely the ones in $\mathfrak{S}$ (compare DUNFORD and SCHWARTZ [1], pp. 418—421).

In the case $\mathscr{L}$ is already a locally convex linear topological (Hausdorff) space to start with and $\mathscr{L}^*$ is its "conjugate" space, that is, the set (necessarily total) of all linear functionals on $\mathscr{L}$ continuous in its topology, then one may still consider the new $\mathscr{L}^*$-topology in $\mathscr{L}$. The last is obviously weaker and often strictly weaker than the originally given topology in $\mathscr{L}$. By what was said already the linear functionals on $\mathscr{L}$ continuous in this new ($\mathscr{L}^*$-topology) are precisely those in $\mathscr{L}^*$.

The preceding remarks should be kept in mind in connection with some of the arguments in the remainder of this section. As the setting for the discussion we choose the linear space $\mathfrak{A}$ of all operators on an infinite-dimensional space $\mathfrak{H}$. There are various operator-topologies which can be introduced in $\mathfrak{A}$. Of special interest to us will be the one which VON NEUMANN [5] calls the *strongest:* A base at $B_0$ is given by the sets

$$\{B \in \mathfrak{A}:\ \Sigma_i \|(B - B_0) f_i\|^2 < \varepsilon\}$$

where $\varepsilon > 0$ and $\{f_i\}$ is an arbitrary sequence of vectors in $\mathfrak{H}$ such that $\Sigma_i \|f_i\|^2 < +\infty$. With this topology $\mathfrak{A}$ is a locally convex linear topological space.

It is easy to see that the strongest topology in $\mathfrak{A}$ is strictly weaker than the uniform topology. We next show that the $(\tau c)$-topology in $\mathfrak{A}$ is strictly weaker than the strongest. Nonetheless the conjugate space of $\mathfrak{A}$ with the strongest topology consists precisely of all the linear functionals in $(\tau c)$, hence coincides with the conjugate space of $\mathfrak{A}$ with the $(\tau c)$-topology. These results (Lemma 3 and Theorem 6) are due to DIXMIER [1].

We repeat first that the $(\tau c)$ topology in $\mathfrak{A}$ is the weakest topology determined by the linear functionals of the form $\mathscr{F}(B) = t(TB)$ where $T$ is fixed in $(\tau c)$ and $B$ varies in $\mathfrak{A}$. Also $(\tau c)$ is total since $t((\varphi \otimes \overline{\psi})\, B) = (B\varphi, \psi) = 0$ for all pairs of vectors $\varphi$ and $\psi$, implies $B = 0$.

Lemma 3. *The $(\tau c)$-topology in $\mathfrak{A}$ is strictly weaker than the strongest topology.*

*Proof.* We first prove that a neighborhood of 0 in the $(\tau c)$-topology contains a neighborhood of 0 in the strongest topology. For this it is sufficient to show that any set of the form

$$\{B \in \mathfrak{A}:\ |t(TB)| < \varepsilon\}$$

for some fixed $T$ in $(\tau c)$ and $\varepsilon > 0$, contains a neighborhood of 0 in the strongest topology. This is easy. We write $T$ in the polar form $T = \Sigma_i \lambda_i \varphi_i \otimes \overline{\psi}_i$ with $\Sigma_i \lambda_i < +\infty$. Then $\Sigma_i \|\lambda_i^{\frac{1}{2}} \varphi_i\|^2 = \Sigma_i \lambda_i$ and thus,

$$\left\{B \in \mathfrak{A} : \Sigma_i \|B(\lambda_i^{\frac{1}{2}} \varphi_i)\|^2 < \frac{\varepsilon^2}{\Sigma_i \lambda_i}\right\}$$

is a neighborhood of 0 in the strongest topology which satisfies our requirements. In fact, for any $B$ in the last set we have

$$|t(TB)| = |\Sigma_i \lambda_i (B\varphi_i, \psi_i)| \leqq \Sigma_i \lambda_i \|B\varphi_i\|$$
$$= \Sigma_i \lambda_i^{\frac{1}{2}} \|\lambda_i^{\frac{1}{2}} B\varphi_i\| \leqq (\Sigma_i \lambda_i)^{\frac{1}{2}} (\Sigma_i \|B(\lambda_i^{\frac{1}{2}} \varphi_i)\|^2)^{\frac{1}{2}} < \varepsilon .$$

To verify that the $(\tau c)$-topology in $\mathfrak{A}$ is indeed strictly weaker than the strongest topology, choose an orthonormal sequence $\{f_i\}$ in $\mathfrak{H}$ and form the sequence of operators $f_i \otimes \bar{f}_1$ ; $i = 1, 2, \ldots$

The last converges to 0 in the $(\tau c)$-topology since $f_i \rightharpoonup 0$ and thus $t(T(f_i \otimes \bar{f}_1)) = (Tf_i, f_1) \to 0$. On the other hand, $\{B : \|Bf_1\|^2 < \frac{1}{2}\}$ is a neighborhood of 0 in the strongest topology which contains none of the operators in our sequence, since

$$\|(f_i \otimes \bar{f}_1) f_1\|^2 = \|(f_1, f_1) f_i\|^2 = 1 .$$

Theorem 6. *The linear functionals on $\mathfrak{A}$ which are continuous in the strongest topology are precisely those in* $(\tau c)$.

*Proof.* Lemma 3 implies that the linear functionals on $\mathfrak{A}$ belonging to $(\tau c)$ are also continuous in the strongest topology. Conversely. Any $\mathscr{F}$ continuous in the strongest topology is of course also continuous in the uniform topology. Hence it may be represented in one and only one way in the form $\mathscr{F} = \mathscr{F}_1 + \mathscr{F}_2$ with $\mathscr{F}_1 \in (\tau c)$ and $\mathscr{F}_2 \in \mathfrak{C}^\perp$. By Lemma 3, $\mathscr{F}_2 = \mathscr{F} - \mathscr{F}_1$ is also continuous in the strongest topology. On the other hand $\mathscr{F}_2(X) = 0$ for all operators $X$ of finite rank, and these operators form an everywhere dense set in the strongest topology. Thus, $\mathscr{F}_2$ is identically 0 on $\mathfrak{A}$. Therefore, $\mathscr{F} = \mathscr{F}_1$.

Remark. Introducing in $\mathfrak{A}$ the bound of an operator as its norm (uniform topology) one obtains $\mathfrak{B}$. The set $\mathfrak{B}^*$ of all continuous linear functionals on $\mathfrak{B}$, determines of course the $\mathfrak{B}^*$-topology in $\mathfrak{A}$. It is easy to see that the last is strictly weaker than the uniform topology and is not comparable with the strongest topology in $\mathfrak{A}$.

We add the following comments. The *weak operator topology* introduced in $\mathfrak{A}$ by VON NEUMANN [7] may be described as follows: A base at $B_0$ is given by all sets of the form $\{B \in \mathfrak{A} : |((B - B_0) f_i, g_i)| < \varepsilon$ for $i = 1, \ldots, n\}$ where $\varepsilon > 0$ and $f_1, \ldots, f_n, g_1, \ldots, g_n$ are any $2n$ vectors in $\mathfrak{H}$; $n = 1, 2, \ldots$ It is obviously the weakest topology in $\mathfrak{A}$ which assures continuity for all the linear functionals of the form $\mathscr{F}(B) = t(BX)$ where $X$ is an operator on $\mathfrak{H}$ of finite rank. With this topology $\mathfrak{A}$ is thus a locally convex linear topological space.

The various topologies introduced in $\mathfrak{A}$ have led one to consider algebras of operators on $\mathfrak{H}$, which as subsets of $\mathfrak{A}$ are also closed with respect to one or another topology. Accordingly a *$(C^*)$-algebra* and a *$(W^*)$-algebra* is by definition a self-adjoint complex algebra of operators on a Hilbert space, closed in the uniform and weak operator topology respectively. Thus, a $(W^*)$-algebra is necessarily also a $(C^*)$-algebra. We may also add that the $(B^*)$-algebras[1] considered in the literature (see RICKART [2]) coincide with the $(C^*)$-algebras. This last result is due to FUKAMIYA [1]. A modification of the argument employed in proving Theorems 1 and 2, proves that *the second conjugate space of a $(C^*)$-algebra considered as a Banach space may be identified with a $(W^*)$-algebra considered as a Banach space.* This was first announced by SHERMAN [1] but the details of the proof were supplied by TAKEDA [1].

Finally, we mention that also $\mathfrak{B}/\mathfrak{C}$ has received some attention in the literature. GELFAND and NEUMARK [1] mention in passing, that *if $\mathfrak{T}$ is a closed ideal in a $(C^*)$-algebra $\mathscr{A}$, then $\mathscr{A}/\mathfrak{T}$ may be again identified with a $(C^*)$-algebra.* The proof of this assertion was finally furnished by KAPLANSKY [1] who makes use of an argument due to I. SEGAL [1]. In particular we have the following corollary first stated by CALKIN [1]: *Let $\mathfrak{B}$ be the Banach algebra of all operators on a Hilbert space $\mathfrak{H}$ and $\mathfrak{C}$ the ideal of all completely continuous operators on $\mathfrak{H}$. Then $\mathfrak{B}/\mathfrak{C}$ is isomorphic in a norm and *-preserving manner, with a uniformly closed self-adjoint Banach algebra of operators on a suitable Hilbert space.*

# V. Norm ideals

## 1. Crossnorms and norm ideals

Again $\mathfrak{A}$ stands for the algebra of all operators on $\mathfrak{H}$ and $\mathfrak{R}$ denotes its subalgebra of all operators of finite rank.

Definition 1. Let $\mathfrak{T}$ be an ideal in $\mathfrak{A}$. A norm on $\mathfrak{T}$ is a real valued function $\alpha = \alpha(A)$ of operators $A \in \mathfrak{T}$ satisfying the following conditions:

(i) $\alpha(A) \geqq 0$; $\alpha(A) = 0$ implies $A = 0$.
(ii) $\alpha(cA) = |c|\,\alpha(A)$ for any complex number $c$.
(iii) $\alpha(A + B) \leqq \alpha(A) + \alpha(B)$.

The bound $\|A\|$ of an operator $A$ is of course a norm. In general, however, a norm does not coincide with the bound. The norm $\alpha$ is a *crossnorm* if it also possesses the "cross-property", that is, if

(iv) $\alpha(A) = \|A\|$ for all operators $A$ of rank 1.

We term $\alpha$ *unitarily invariant* if

(v) $\alpha(UAV^*) = \alpha(A)$ for $A$ in $\mathfrak{T}$ and any pair $U, V$ of unitary operators on $\mathfrak{H}$.

---

[1] The abstract definition of a $(B^*)$-algebra is that of a Banach algebra with an involution $f \to f^*$ and whose norm also satisfies the condition $\|f^* f\| = \|f\|^2$.

We define $\alpha$ as *uniform* if

(vi) $\alpha(XAY) \leqq \|X\| \, \|Y\| \, \alpha(A)$ for $A$ in $\mathfrak{T}$ and any pair $X, Y$ of operators on $\mathfrak{H}$.

A crossnorm $\alpha$ is termed unitarily invariant or uniform if in addition to properties (i)—(iv), it also satisfies (v) or (vi).

Lemma 1. *Every uniform crossnorm $\alpha$ is necessarily unitarily invariant.*

*Proof.* Uniformity for $\alpha$ implies

$$\alpha(UAV^*) \leqq \|U\| \, \|V^*\| \, \alpha(A) = \alpha(A) .$$

On the other hand,

$$\alpha(A) = \alpha(U^*UAV^*V) \leqq \|U^*\| \, \|V\| \, \alpha(UAV^*) = \alpha(UAV^*) .$$

Remark. We shall prove later a converse statement, namely that every unitarily invariant crossnorm on $\mathfrak{R}$ is necessarily uniform.

Definition 2. Let $\alpha$ be a crossnorm on $\mathfrak{R}$. For an operator $A \in \mathfrak{R}$ define

$$\alpha'(A) = \sup_{0 \neq X \in \mathfrak{R}} \frac{|t(XA)|}{\alpha(X)} .$$

Thus, $\alpha'$ is again a non-negative function on $\mathfrak{R}$ termed the *associate* with $\alpha$.

Theorem 1. *Let $\alpha$ be a crossnorm on $\mathfrak{R}$. Then $\alpha'$ is also a crossnorm on $\mathfrak{R}$ if and only if $\alpha(X) \geqq \|X\|$ for all $X \in \mathfrak{R}$.*

*Proof.* Let $\alpha$ be any crossnorm. For any pair of non-zero vector $\varphi$ and $\psi$ we have

$$\alpha'(\varphi \otimes \overline{\psi}) \geqq \frac{|t(\psi \otimes \overline{\varphi})\,(\varphi \otimes \overline{\psi})|}{\alpha(\psi \otimes \overline{\varphi})} = \|\varphi\| \, \|\psi\| .$$

Assuming in addition $\alpha(X) \geqq \|X\|$ for all $X \in \mathfrak{R}$, then

$$\frac{|t(X(\varphi \otimes \overline{\psi}))|}{\alpha(X)} \leqq \frac{|(X\varphi, \psi)|}{\|X\|} \leqq \|\varphi\| \, \|\psi\|$$

and therefore also $\alpha'(\varphi \otimes \overline{\psi}) \leqq \|\varphi\| \, \|\psi\|$. Thus, $\alpha'$ has the cross-property. It follows that $\alpha'(A)$ is finite for every $A \in \mathfrak{R}$. That $\alpha'$ is also a norm is immediate.

Assume on the other hand that for some $X \in \mathfrak{R}$ we have $\alpha(X) < \|X\|$. Choosing a pair of vectors $\varphi$ and $\psi$ both of norm 1 such that $\alpha(X) < |(X\varphi, \psi)|$ one gets

$$\alpha'(\varphi \otimes \overline{\psi}) \geqq \frac{|(X\varphi, \psi)|}{\alpha(X)} > 1 .$$

Thus $\alpha'$ is not a crossnorm.

Corollary. *If both $\alpha$ and $\alpha'$ are crossnorms, the same is also true for $\alpha'', \alpha''', \ldots$*

*Proof.* If $\alpha$ is a crossnorm, then for $A \in \mathfrak{R}$ one has

$$\alpha'(A) = \sup_{0 \neq X \in \mathfrak{R}} \frac{|t(XA)|}{\alpha(X)} \geqq \sup_{\varphi \neq 0;\, \psi \neq 0} \frac{|(A\varphi, \psi)|}{\|\varphi\| \, \|\psi\|} = \|A\| .$$

An application of the preceding theorem concludes the proof.

It is interesting that *even for a finite $n \geqq 2$, we can construct crossnorms on $\mathfrak{H}_n$ whose associates are not crossnorms.* Such crossnorms must not be unitarily invariant. To do this, we make use of a minimum property of the crossnorm $\tau(X)$.

Theorem 2. *Let $X$ be an operator of rank 1 on a finite-dimensional space $\mathfrak{H}_n$; $n \geqq 2$. Then for every complex number $c$ we have*

$$\tau(X - cI) \geqq \tau(X) = \|X\| .$$

*Proof.* We first consider the case $n = 2$. Let $\varphi_2$ be the vector of norm 1 determining the range of $X$, and $\varphi_1$, $\varphi_2$ form a basis in $\mathfrak{H}_2$. Then $X$ determines a matrix of the form

$$\begin{pmatrix} 0, & 0 \\ a, & b \end{pmatrix}$$

Replacing if necessary $X$ by $\theta X$ — where $\theta$ is complex with $|\theta| = 1$ — we may assume that $b \geqq 0$. Since $\varphi_1$ may be replaced by $\theta_1 \varphi_1$ — where $\theta_1$ is complex with $|\theta_1| = 1$ — it may be also assumed $a \geqq 0$. The operator $X - cI$ determines then the matrix

$$\begin{pmatrix} -c, & 0 \\ a, & b - c \end{pmatrix}$$

with $a \geqq 0$, $b \geqq 0$ and $c$ complex. It is clear that the matrix corresponding to the operator $(X - cI)^* (X - cI) = [X - cI]^2$ has as its trace the quantity $a^2 + b^2 + 2|c|^2 - 2b\mathscr{R}c$ and as its determinant the value $|c|^2 |b - c|^2$.

Let $\lambda_1$, $\lambda_2$ represent the proper values of $[X - cI]$. Then $\lambda_1^2$, $\lambda_2^2$ are the proper values of $(X - cI)^* (X - cI)$ and thus,

$$\lambda_1^2 + \lambda_2^2 = a^2 + b^2 + 2|c|^2 - 2b\mathscr{R}c$$

$$\lambda_1 \lambda_2 = |c|\, |b - c| .$$

Therefore,

$$\tau(X - cI) = \lambda_1 + \lambda_2 = \sqrt{\lambda_1^2 + \lambda_2^2 + 2\lambda_1\lambda_2} =$$

$$= \sqrt{a^2 + b^2 + 2|c|^2 - 2b\mathscr{R}c + 2|c|\,|b - c|} .$$

Putting $c = 0$, we get

$$\tau(X) = \sqrt{a^2 + b^2} .$$

The inequality we set out to prove assumes therefore the following form:

$$\sqrt{a^2 + b^2 + 2|c|^2 - 2b\mathscr{R}c + 2|c|\,|b - c|} \geqq \sqrt{a^2 + b^2}$$

that is,

$$|c|\,|b - c| \geqq b\mathscr{R}c - |c|^2 .$$

The last however is obvious since,

$$|c|\,|b - c| \geqq |c|\,|b| - |c|^2 \geqq b\mathscr{R}c - |c|^2 .$$

Next consider the case $n > 2$. Since $X$ is of rank 1 it is of the form $\varphi \otimes \overline{\psi}$. Let $\mathfrak{H}'$ be a two-dimensional subspace containing $\varphi$ and $\psi$.

Then, $X$ is completely reduced by $\mathfrak{H}'$, that is, $X$ (and of course I also) may be considered as an operator on $\mathfrak{H}'$ and as an operator on $\mathfrak{H}_n \ominus \mathfrak{H}'$. In $\mathfrak{H}'$, the operator $X$ still has rank 1, while in $\mathfrak{H}_n \ominus \mathfrak{H}'$ the $X$ is identically zero. Therefore,

$$\tau(X - cI) = \tau_{\mathfrak{H}'}(X - cI) + \tau_{\mathfrak{H}_n \ominus \mathfrak{H}'}(X - cI) \geqq \tau_{\mathfrak{H}'}(X - cI)\,,$$
$$\tau(X) = \tau_{\mathfrak{H}'}(X) + \tau_{\mathfrak{H}_n \ominus \mathfrak{H}'}(X) = \tau_{\mathfrak{H}'}(X)\,.$$

Thus, it is sufficient to prove

$$\tau_{\mathfrak{H}'}(X - cI) \geqq \tau_{\mathfrak{H}'}(X)\,.$$

In other words, the whole problem may be considered entirely within $\mathfrak{H}'$. The first part of our proof furnishes the desired conclusion.

Now for the promised construction: Let $0 < \varepsilon < 1$. For $X$ in $\mathfrak{R}_n$ define

$$\alpha(X) = \varepsilon\,\|X\| + (1 - \varepsilon)\ \inf \tau(X - cI)$$

where the infimum is taken over all complex numbers $c$. We verify at once that $\alpha$ is a norm. The previous result proves that $\alpha$ has also the cross-property. Since $\alpha(I) = \varepsilon < 1$, Theorem 1 implies that $\alpha'$ is not a crossnorm.

Remark. In the case $\mathfrak{H}$ is infinite dimensional then $I$ does not belong to $(\tau c)$. It is however not difficult to take care also of this case by a modification of the above procedure in the course of which $I$ is replaced by a projection on $\mathfrak{H}$ whose range is at least four-dimensional.

Definition 3. An ideal $\mathfrak{T} \subset \mathfrak{A}$ is termed a *norm ideal* if on it there is defined a uniform crossnorm with respect to which $\mathfrak{T}$ is also complete. A norm ideal $\mathfrak{T}$ is *minimal* if none of its proper subspaces is also a norm ideal. $\mathfrak{T}$ is a norm ideal of completely continuous operators if every one of its elements is a completely continuous operator.

Thus, with every operator $A$ in a norm ideal $\mathfrak{T} \subset \mathfrak{A}$ there is associated its norm $\alpha(A)$ and of course also its bound $\|A\|$; for operators which are not of rank $\leqq 1$ the two are in general different. Clearly, the Schmidt-class with the norm $\sigma(A)$, the trace-class wirh the norm $\tau(A)$, the space of all completely continuous operators with $\|A\|$ as the norm are examples of minimal norm ideals of completely continous operators.

We are about to "determine" all minimal norm ideals as well as characterize their conjugate spaces which as we shall see, again may be interpreted as norm ideals of operators. The quotation marks around the word "determine" are there for the following reason: Actually the construction of the minimal norm ideals is shifted to the construction of the symmetric gauge functions since they generate each other. This adds a different flavor to our problem since the symmetric gauge functions are comparatively easy to deal with. We shall also see that all the operators in a minimal norm ideal are completely continuous.

## 2. A maximum problem for $\mathfrak{H}_n$

We consider a maximum problem for the trace $t(A)$ on $\mathfrak{R}_n$. Its exact formulation and solution is stated in Theorem 3 below. This result was first derived by VON NEUMANN [6]. On it we base some of our later discussion.

The setting is thus a fixed finite say $n$-dimensional Hilbert space $\mathfrak{H}_n$; all operators considered below (as well as the assertions involving them) are assumed to be defined on (or refer to) this $\mathfrak{H}_n$.

Lemma 2. *If $A$ is fixed and $U$ varies over the set of all unitary operators, the expression $\mathscr{R}t(UA)$ assumes a maximum and its value is $\tau(A)$. Also $U_0$ is maximal, that is, $\mathscr{R}t(U_0A) \geqq \mathscr{R}t(UA)$ for all unitary $U$ if and only if $U_0A \geqq 0$.*

*Proof.* Clearly, there are unitary $U$ for which $UA = [A]$. An application of Theorem III.6 (the first two corollaries) concludes the proof.

Lemma 3. *The expression $\mathscr{R}t(UAVB)$ when $A$ and $B$ are fixed and $U, V$ vary over the set of all pairs of unitary operators, assumes a maximum. If the pair $U_0$, $V_0$ is maximal, then both $(U_0AV_0)B$ and $B(U_0AV_0)$ are positive.*

*Proof.* Consider the algebra $\mathscr{A}$ of all $n \times n$ complex matrices $(a_{ij})$ with the norm $(\Sigma_{ij} |a_{ij}|^2)^{\frac{1}{2}}$. Relative to a fixed basis $\varphi_1, \ldots, \varphi_n$ in $\mathfrak{H}_n$ the relations $A\varphi_j = a_{1j}\varphi_1 + \cdots + a_{nj}\varphi_n$ for $j = 1, \ldots, n$ establish a norm preserving isomorphism between $\mathscr{A}$ and the $(\sigma c)$ of operators on $\mathfrak{H}_n$, that is, the algebra of all operators $A$ on $\mathfrak{H}_n$ with $\sigma(A)$ representing the norm of $A$. In this isomorphism the unitary matrices correspond to the unitary operators.

Clearly $\mathscr{R}t(UAVB)$ is a real valued continuous function on the closed and compact set of all pairs of unitary operators (matrices). Consequently it assumes a maximum. If $U_0$ and $V_0$ is a maximal pair, then

$$\mathscr{R}t(U_0AV_0B) \geqq \mathscr{R}t(UU_0AV_0VB)$$

for every pair $U$, $V$ of unitary operators. An application of Theorem III.6 (the last corollary) concludes the proof.

Lemma 4. *Let $AB = BA = \lambda I$ with $\lambda \geqq 0$. Then there is a unitary $U$ for which both $AU$ and $U^{-1}B$ are positive.*

*Proof.* Consider first the case $\lambda > 0$: Then $A^{-1}$ exists and equals $\frac{1}{\lambda} B$. We choose $U$ so that $AU \geqq 0$. Then $(AU)^{-1} = \frac{1}{\lambda}U^{-1}B$ is also positive and therefore $U^{-1}B \geqq 0$.

The case $\lambda = 0$: Choose a unitary $V$ so that $AV \geqq 0$. If $\mathfrak{R}$ and $\mathfrak{N}$ stand for the range and nullspace of $AV$ we have

$$\mathfrak{R} \oplus \mathfrak{N} = \mathfrak{H}_n \, .$$

Furthermore, since $AV$ and $V^{-1}B$ satisfy the obvious relations

$$AV \cdot V^{-1}B = AB = 0\,,$$
$$V^{-1}B \cdot AV = V^{-1} \cdot BA \cdot V = 0$$

it follows that the range of one is included in the nullspace of the other. Thus the pair $\{\mathfrak{R}, \mathfrak{N}\}$ reduces both operators $AV$ and $V^{-1}B$; $AV$ is 0 on $\mathfrak{N}$ while $V^{-1}B$ is 0 on $\mathfrak{R}$. Choose a unitary $W_0$ on $\mathfrak{N}$ for which $W_0^{-1}V^{-1}B \geqq 0$ on $\mathfrak{N}$. We extend then $W_0$ to a unitary operator $W$ on $\mathfrak{H}_n$ requiring that $W$ coincide with the identity operator on $\mathfrak{R}$. Clearly, $U = VW$ satisfies our requirements: $U^{-1}B$ being equal $W_0^{-1}V^{-1}B$ on $\mathfrak{N}$ and 0 on $\mathfrak{R}$, is obviously $\geqq 0$ on $\mathfrak{H}_n$. Similarly, $AU \geqq 0$ on $\mathfrak{H}_n$ since it coincides with $AV$ on $\mathfrak{R}$ and with 0 on $\mathfrak{N}$.

Lemma 5. *Let $AB \geqq 0$ and $BA \geqq 0$. Then for some unitary $U$ the operators $A' = AU$, $B' = U^{-1}B$ satisfy the following conditions:*

$$A' \geqq 0 \text{ and } B' \geqq 0\,,$$
$$AB = A'B' = B'A'\,.$$

*Proof* [1]. Let $\lambda_1, \ldots, \lambda_p$ be all the different proper values of $AB$ hence also of $BA$ and let $\mathfrak{M}_k$ and $\mathfrak{N}_k$ be the characteristic subspaces for $\lambda_k$ when considered as proper value of $AB$ and of $BA$ respectively. $\mathfrak{M}_k$ and $\mathfrak{N}_k$ have the same dimension and

$$\mathfrak{M}_1 \oplus \cdots \oplus \mathfrak{M}_p = \mathfrak{N}_1 \oplus \cdots \oplus \mathfrak{N}_p = \mathfrak{H}_n\,.$$

Thus, there is a unitary $V$ which maps $\mathfrak{M}_k$ onto $\mathfrak{N}_k$ for $k = 1, \ldots, p$. Put $\tilde{A} = AV$, $\tilde{B} = V^{-1}B$. The operators $\tilde{A}\tilde{B} = AB$ and $\tilde{B}\tilde{A} = V^{-1}BAV$ are also $\geqq 0$, have the same proper values $\lambda_1, \ldots, \lambda_p$ and $\mathfrak{M}_k$ is the characteristic subspace for $\lambda_k$ both when it is considered as a proper value of $\tilde{A}\tilde{B}$ and also of $\tilde{B}\tilde{A}$. This means that on each $\mathfrak{M}_k$

$$AB = \tilde{A}\tilde{B} = \tilde{B}\tilde{A} = \lambda_k I\,.$$

[1] Let $A$ and $B$ be two operators defined perhaps on an infinite-dimensional space. Then $\lambda \neq 0$ is a proper value for $AB$ of finite multiplicity if and only if it is also a proper value for $BA$ and with the same multiplicity: Assume that $\lambda \neq 0$ is a proper value for $AB$, and $\varphi_1, \ldots, \varphi_p$ are linearly independent — corresponding to $\lambda$ — proper vectors for $AB$. Then, the vectors $B\varphi_1, \ldots, B\varphi_p$ are also linearly independent since $\Sigma_{i=1}^{p} a_i B\varphi_i = 0$ implies

$$\Sigma_{i=1}^{p} a_i \varphi_i = \Sigma_{i=1}^{p} a_i \frac{1}{\lambda} AB\varphi_i = \frac{1}{\lambda} A(\Sigma_{i=1}^{p} a_i B\varphi_i) = 0\,.$$

Moreover, we have,

$$BA(B\varphi_i) = B(AB\varphi_i) = B(\lambda\varphi_i) = \lambda B\varphi_i\,.$$

It follows that $\lambda$ is also a proper value for $BA$ of multiplicity at least $p$. Interchanging the rôles of $AB$ and $BA$ we obtain the desired conclusion.

In particular, if $AB$ and $BA$ are Hermitean operators on a finite-dimensional space $\mathfrak{H}_n$, then $n$ minus the sum of the multiplicities of the non-zero proper values, represents the multiplicity of the proper value 0 for both $AB$ and $BA$.

Now, $\varphi \in \mathfrak{M}_k$ implies $\tilde{B}\tilde{A}\varphi = \lambda_k \varphi$ and therefore $\tilde{A}\tilde{B}(\tilde{A}\varphi) = \tilde{A}(\tilde{B}\tilde{A}\varphi) = \lambda_k \tilde{A}\varphi$. This means $\varphi \in \mathfrak{M}_k$ implies $\tilde{A}\varphi$ is in $\mathfrak{M}_k$. Similarly $\varphi \in \mathfrak{M}_k$ implies that $\tilde{B}\varphi$ is in $\mathfrak{M}_k$. Thus each $\mathfrak{M}_k$ is invariant under both $\tilde{A}$ and $\tilde{B}$. On $\mathfrak{M}_k$ we have $\tilde{A}\tilde{B} = \tilde{B}\tilde{A} = \lambda_k I$. Thus Lemma 4 furnishes a unitary $W_k$ on $\mathfrak{M}_k$, for which $\tilde{A}W_k$ and $W_k^{-1}\tilde{B}$ are $\geqq 0$. The operator $W$ on $\mathfrak{H}_n$ determined by the requirements $W = W_k$ on $\mathfrak{M}_k$ for $k = 1, \ldots, p$ is clearly unitary. Moreover, $\tilde{A}W = AVW$ and $W^{-1}\tilde{B} = W^{-1}V^{-1}B$ being $\geqq 0$ on each $\mathfrak{M}_k$ are also $\geqq 0$ on $\mathfrak{H}_n$. Therefore, the unitary operator $U = VW$ satisfies our requirements.

Theorem 3. *If $A$ and $B$ are fixed and $U, V$ vary over all pairs of unitary operators, then the expression $\mathfrak{R}t(UAVB)$ assumes a maximum. The last is given by*

$$\Sigma_{i=1}^n \lambda_i \mu_i$$

*where* $\lambda_1 \geqq \cdots \geqq \lambda_n$ and $\mu_1 \geqq \cdots \geqq \mu_n$ *are the proper values of* $[A]$ *and* $[B]$ *respectively.*

*Proof.* By Lemma 3 we can find a maximal pair $U_0$, $V_0$. For such a pair both $U_0AV_0 \cdot B$ and $B \cdot U_0AV_0$ are $\geqq 0$. Lemma 5 permits then to construct a unitary operator $W$ such that $\tilde{A} = (U_0AV_0)W$ and $\tilde{B} = W^{-1}B$ are positive commutative operators; we have of course $U_0AV_0B = \tilde{A}\tilde{B}$.

By Theorem I.6 (corollary) there is a basis $\varphi_1, \ldots, \varphi_n$ all of whose elements are proper vectors for both $\tilde{A}$ and $\tilde{B}$. Let $a_i$ and $b_i$ be the proper values corresponding to $\varphi_i$ for $\tilde{A}$ and $\tilde{B}$ respectively. Clearly, $a_i \geqq 0$ and $b_i \geqq 0$. Of course, we have also $\tilde{A}\tilde{B}\varphi_i = a_i b_i \varphi_i$. Therefore,

$$t(U_0AV_0B) = t(\tilde{A}\tilde{B}) = \Sigma_{i=1}^n a_i b_i\,.$$

Since $B^*B = \tilde{B}^2$, the $\mu_1^2, \ldots, \mu_n^2$ form a permutation of the $b_1^2, \ldots, b_n^2$, and therefore the $\mu_1, \ldots, \mu_n$ form a permutation of the $b_1, \ldots, b_n$. Similarly, $A^*A = V_0W\tilde{A}^2W^*V_0^*$ implies that the $\lambda_1, \ldots, \lambda_n$ form a permutation of the $a_1, \ldots, a_n$.

Since the $\varphi_i$, $a_i$, $b_i$ are determined up to the same permutation on their indices, we may of course assume that $a_i = \lambda_i$. We then have

$$t(U_0AV_0B) = \Sigma_{i=1}^n \lambda_i \mu_{p(i)}$$

where $\{p(1), \ldots, p(n)\}$ is some permutation on $\{1, \ldots, n\}$. On the other hand it is easy to see that if $\{q(1), \ldots, q(n)\}$ is any given permutation on $\{1, \ldots, n\}$, a unitary pair $U, V$ can be chosen so that

$$t(UAVB) = \Sigma_{i=1}^n \lambda_i \mu_{q(i)}\,.$$

Thus, the maximum we set out to compute coincides with the maximum of all the sums $\Sigma_{i=1}^n \lambda_i \mu_{p(i)}$ corresponding to all the possible permutations $\{p(1), \ldots, p(n)\}$ on $\{1, \ldots, n\}$. Since, $\lambda_1 \geqq \cdots \geqq \lambda_n$ and $\mu_1 \geqq \ldots \geqq \mu_n$, the last maximum is easily seen to be $\Sigma_{i=1}^n \lambda_i \mu_i$.

## 3. Symmetric gauge functions on $\mathfrak{L}_n$

Let $\mathfrak{L}_n$ stand for the $n$-dimensional space of $n$-tuples of real numbers $u = (u_1, \ldots, u_n)$ referred to below as points. If $x = (x_1, \ldots, x_n)$ and $y = (y_1, \ldots, y_n)$ are two points, we write $(x, y)$ for the sum $\Sigma_{i=1}^n x_i y_i$. For a fixed point $u \neq 0$ and a real constant $c$, the equation $(u, x) = c$ defines a plane in $\mathfrak{L}_n$; the inequalities $(u, x) \geqq c$ and $(u, x) \leqq c$ define the two closed half-spaces associated with the above plane.

A closed and bounded convex set in $\mathfrak{L}_n$ with a non-empty interior is termed a convex body. The convex hull $\tilde{\mathfrak{S}}$ of a closed set $\mathfrak{S}$ is defined as the intersection of all closed half-spaces containing $\mathfrak{S}$; whenever no such half-space exists, we define $\tilde{\mathfrak{S}}$ to be the whole space.

The theorem which follows is one of the basic in the theory of convex sets. Its proof is quite elementary (see for instance BONNESEN and FENCHEL [1], p. 5): *The convex hull $\tilde{\mathfrak{S}}$ of a closed and bounded set $\mathfrak{S}$ is the intersection of all closed convex sets containing $\mathfrak{S}$. Thus, if $\mathfrak{S}$ is closed, bounded and convex (and in particular a convex body) then $\tilde{\mathfrak{S}} = \mathfrak{S}$.*

Definition 4. A real valued function $\Phi(u) = \Phi(u_1, \ldots, u_n)$ on $\mathfrak{L}_n$ is termed a *gauge function*[1] if it satisfies the following conditions:

(i) $\Phi(u_1, \ldots, u_n) > 0$ unless $u_1 = \cdots = u_n = 0$.

(ii) $\Phi(cu_1, \ldots, cu_n) = c\,\Phi(u_1, \ldots, u_n)$ for any constant $c \geqq 0$.

(iii) $\Phi(u_1 + u_1', \ldots, u_n + u_n') \leqq \Phi(u_1, \ldots, u_n) + \Phi(u_1', \ldots, u_n')$.

$\Phi$ is a *symmetric* gauge function if in addition to (i), (ii) and (iii) it also satisfies the following condition

(iv) $\Phi(u_1, \ldots, u_n) = \Phi(\varepsilon_1 u_{p(1)}, \ldots, \varepsilon_n u_{p(n)})$

if $\varepsilon_i = \pm 1$ and $p(1), \ldots, p(n)$ stands for any permutation on $1, \ldots, n$.

To simplify our formulae we shall always assume that a symmetric $\Phi$ also satisfies the following condition:

(v) $\Phi(1, 0, \ldots, 0) = 1$.

Lemma 6. *Let $\Phi(u_1, \ldots, u_n)$ denote a symmetric gauge function on $\mathfrak{L}_n$. Then, for $0 \leqq p_i \leqq 1$ we have,*

$$\Phi(p_1 u_1, \ldots, p_n u_n) \leqq \Phi(u_1, \ldots, u_n)\ .$$

*Proof.* By virtue of (iv) we may suppose that all the $u_i$'s are $\geqq 0$. Moreover, it is obviously sufficient to establish the last relationship when $p_i \neq 1$ occurs only for one $i$, that is,

$$\Phi(u_1, \ldots, u_{i-1}, p u_i, u_{i+1}, \ldots, u_n) \leqq$$
$$\leqq \Phi(u_1, \ldots, u_{i-1}, u_i, u_{i+1}, \ldots, u_n)$$

[1] Since $u = (u_1, \ldots, u_n)$ we really should write $\Phi(u) = \Phi((u_1, \ldots, u_n))$.

for $0 \leqq p < 1$. The last readily follows from the following simple and direct calculation:

$$\begin{aligned}
&\Phi(u_1, \ldots, p u_i, \ldots, u_n) = \\
&\Phi\left(\frac{1+p}{2} u_1 + \frac{1-p}{2} u_1, \ldots, \frac{1+p}{2} u_i + \frac{1-p}{2}(-u_i), \ldots,\right. \\
&\qquad \left.\frac{1+p}{2} u_n + \frac{1-p}{2} u_n\right) \leqq \Phi\left(\frac{1+p}{2} u_1, \ldots, \frac{1+p}{2} u_i, \ldots, \frac{1+p}{2} u_n\right) + \\
&\qquad\qquad + \Phi\left(\frac{1-p}{2} u_1, \ldots, \frac{1-p}{2}(-u_i), \ldots, \frac{1-p}{2} u_n\right) = \\
&= \frac{1+p}{2} \Phi(u_1, \ldots, u_i, \ldots, u_n) + \frac{1-p}{2} \Phi(u_1, \ldots, (-u_i), \ldots, u_n) = \\
&= \frac{1+p}{2} \Phi(u_1, \ldots, u_i, \ldots, u_n) + \frac{1-p}{2} \Phi(u_1, \ldots, u_i, \ldots, u_n) = \\
&= \Phi(u_1, \ldots, u_i, \ldots, u_n) .
\end{aligned}$$

Remark. The above proof holds verbatim for any $\Phi$ satisfying merely conditions (i) — (iv) of Definition 4.

Corollary. $|u_i| \leqq |u_i'|$ *for* $i = 1, \ldots, n$ *implies*

$$\Phi(u_1, \ldots, u_n) \leqq \Phi(u_1', \ldots, u_n') .$$

Lemma 7. *Any symmetric gauge function* $\Phi(u_1, \ldots, u_n)$ *on* $\mathfrak{L}_n$ *satisfies the inequality*

$$\max_i |u_i| \leqq \Phi(u_1, \ldots, u_n) .$$

*Proof.* Lemma 6 gives

$$\Phi(0, \ldots, 0, u_i, 0, \ldots, 0) \leqq \Phi(u_1 \ldots, u_{i-1}, u_i, u_{i+1}, \ldots, u_n) .$$

The left side equals $|u_i|$ by (ii), (iv) and (v) of Definition 4. Hence $|u_i| \leqq \leqq \Phi(u_1, \ldots, u_n)$ for $i = 1, \ldots, n$.

Lemma 8. *For a symmetric gauge function* $\Phi(u_1, \ldots, u_n)$ *on* $\mathfrak{L}_n$ *we have,*

$$\Phi(u_1, \ldots, u_n) \leqq \Sigma_{i=1}^n |u_i| .$$

*Proof.* The proof is a simple consequence of conditions (ii), (iii), (iv) and (v) for $\Phi$.

Lemma 9. *A gauge function* $\Phi(u_1, \ldots, u_n)$ *is continuous on* $\mathfrak{L}_n$.

*Proof.* Conditions (ii) and (iii) for $\Phi$ furnish

$$|\Phi(u_1', \ldots, u_n') - \Phi(u_1, \ldots, u_n)| \leqq \Sigma_{i=1}^n |u_i - u_i'| \, m_i$$

where $m_i$ is the maximum of the two values assumed by $\Phi(x_1, \ldots, x_n)$ when $x_i = \pm 1$ and $x_j = 0$ for $j \neq i$.

The *associate* of a given gauge function $\Phi(u_1, \ldots, u_n)$ on $\mathfrak{L}_n$ is defined as follows: For a fixed $n$-tuple $(v_1, \ldots, v_n)$, the expression

$$\frac{u_1 v_1 + \cdots + u_n v_n}{\Phi(u_1, \ldots, u_n)}$$

represents a continuous function on the closed and bounded set of $n$-tuples $(u_1, \ldots, u_n)$ for which $|u_1| + \cdots + |u_n| = 1$. Hence it assumes there a maximum, which we shall denote by $\Psi(v_1, \ldots, v_n)$. The last coincides, of course, with the maximum value assumed by the above expression on the set of all $n$-tuples $(u_1, \ldots, u_n) \neq (0, \ldots, 0)$.

We also remark that $\Psi(v_1, \ldots, v_n) \leqq 1$ if and only if $\Sigma_{i=1}^n u_i v_i \leqq 1$ whenever $\Phi(u_1, \ldots, u_n) \leqq 1$.

The proof of the following two lemmas is immediate:

Lemma 10. *$\Psi(v_1, \ldots, v_n)$ is a gauge function whenever $\Phi(u_1, \ldots, u_n)$ is such. If $\Phi$ is symmetric, the same is true for $\Psi$.*

Lemma 11. $u_1 v_1 + \cdots + u_n v_n \leqq \Phi(u_1, \ldots, u_n) \; \Psi(v_1, \ldots, v_n)$.

Lemma 12. *Let $\Phi$ be a gauge function and $\Psi$ its associate. Then $\Phi$ is also the associate with $\Psi$.*

*Proof.* Let $X$ stand for the associate with $\Psi$. The full unit sphere

$$\mathscr{K} = \{x : \Phi(x) \leqq 1\}$$

is obviously a convex body in $\mathfrak{L}_n$ containing 0 in its interior. By the mentioned above theorem, $\mathscr{K}$ must coincide with its convex hull, that is, any point $u$ which belongs to all half-spaces containing $\mathscr{K}$ necessarily belongs to $\mathscr{K}$.

The equation of a half-space containing 0 in its interior is of the form $(v, x) \leqq 1$, where $v$ is a non-zero fixed point. The theorem referred to above may thus be expressed as follows: If for a fixed point $u$, we have $(v, u) \leqq 1$ for all $v$ for which $\Phi(x) \leqq 1$ implies $(v, x) \leqq 1$, then also $\Phi(u) \leqq 1$. This of course means: If for a fixed point $u$, we have $(v, u) \leqq 1$ whenever $\Psi(v) \leqq 1$, then $\Phi(u) \leqq 1$. This again means, $X(u) \leqq 1$ implies $\Phi(u) \leqq 1$. Thus, $\Phi(u) \leqq X(u)$ for all $u$. On the other hand $X(u) \leqq \Phi(u)$ is an immediate consequence of the definition of $\Psi$ and $X$ for $\Phi$. Thus, $X(u) = \Phi(u)$ for all $u$.

Clearly, a symmetric gauge function $\Phi$ is also a norm on $\mathfrak{L}_n$. Denoting by $\mathfrak{L}_n(\Phi)$ the linear space $\mathfrak{L}_n$ on which there is defined the symmetric gauge function $\Phi$ we may express the following:

*The conjugate space of the normed linear space $\mathfrak{L}_n(\Phi)$ may be characterized as $\mathfrak{L}_n(\Psi)$.*

*Proof.* It is clear that every linear functional $\mathcal{F}$ on $\mathfrak{L}_n(\Phi)$ determines an $n$-tuple of numbers $(v_1, \ldots, v_n)$ such that

(1) $\mathcal{F}(u_1, \ldots, u_n) = u_1 v_1 + \cdots + u_n v_n$ for $(u_1, \ldots, u_n)$ in $\mathfrak{L}_n(\Phi)$.

Furthermore, the bound

$$(2)\ \|\mathcal{F}\| = \max_{(u_1, \ldots, u_n)} \frac{u_1 v_1 + \cdots + u_n v_n}{\Phi(u_1, \ldots, u_n)} = \Psi(v_1, \ldots, v_n).$$

Conversely. Given an $n$-tuple of numbers $(v_1, \ldots, v_n)$ then (1) above, determines a linear functional $\mathcal{F}$ whose bound is given by (2).

## 4. The unitarily invariant crossnorms on $\mathfrak{R}_n$

We denote by $\mathfrak{R}_n$ the linear space of all operators on the (finite) $n$-dimensional space $\mathfrak{H}_n$. The discussion which follows describes the direct connection between the symmetric gauge functions on $\mathfrak{L}_n$ and the unitarily invariant crossnorms on $\mathfrak{R}_n$.

Definition 5. Let $\Phi(u_1, \ldots, u_n)$ be a symmetric gauge function on $\mathfrak{L}_n$ and $\Psi(v_1, \ldots, v_n)$ its associate. For every operator $X$ in $\mathfrak{R}_n$ define

$$\Phi(X) = \Phi(x_1, \ldots, x_n)$$

where $x_1, \ldots, x_n$ are the proper values of $[X]$; each proper value appearing the number of times equal to its multiplicity.

Theorem 4. *Consider $\mathfrak{R}_n$. When $A$ is fixed and $X$ varies subject to the condition $\Phi(X) = 1$, then each of the expressions*

$$\mathcal{R}t(XA), \quad |t(XA)|, \quad |(A, X)|,$$

*assumes a maximum. They all have the same numerical value equal to $\Psi(A)$.*

*Proof.* Let $U$ and $V$ be a pair of unitary operators. Clearly, $X^*X$ and $(UXV)^*(UXV) = V^*X^*XV$ possess the same proper values. It follows that $[X]$ and $[UXV]$ have also the same proper values and therefore

$$\Phi(X) = \Phi(UXV).$$

In particular $\Phi(X) = 1$ if and only if $\Phi(UXV) = 1$. Consequently,

$$\sup_{\Phi(X)=1} \mathcal{R}t(XA) = \sup_{\Phi(X)=1} \left(\max_{U,V} \mathcal{R}t(UXVA)\right).$$

By Theorem 3, the right side represents thus $\sup \Sigma_{i=1}^{n} a_i x_i$ where the $a_1 \geqq \cdots \geqq a_n$ are the fixed proper values of $[A]$ and $x_1, \ldots, x_n$ is any $n$-tuple of real numbers subject to the restrictions $x_1 \geqq \cdots \geqq x_n \geqq 0$ and $\Phi(x_1, \ldots, x_n) = 1$. A little consideration, however, shows that the requirement $x_1 \geqq \cdots \geqq x_n \geqq 0$ may be omitted. In fact, if $x_i < x_j$ for a pair $i < j$, then the interchange of $x_i$ and $x_j$ does not affect $\Phi(x_1, \ldots, x_n) = 1$

and does not decrease $\Sigma_{i=1}^n a_i x_i$; the change being $(a_i x_j + a_j x_i) - (a_i x_i + a_j x_j) = (a_i - a_j)(x_j - x_i) \geqq 0$. Thus, the condition $x_1 \geqq \cdots \geqq x_n \geqq 0$ may be replaced by $x_i \geqq 0$. But even the last requirement may be omitted; if $x_i < 0$, a replacement of $x_i$ by $-x_i$ does not affect $\Phi(x_1, \ldots, x_n) = 1$ and of course does not decrease $\Sigma_{i=1}^n a_i x_i$. Thus, we are really dealing with $\sup \Sigma_{i=1}^n a_i x_i$ where the $x_1, \ldots, x_n$ are subject to the sole restriction $\Phi(x_1, \ldots, x_n) = 1$. That is,

$$\sup_{\Phi(X)=1} \mathscr{R}\, t(XA) = \sup_{\Phi(x_1, \ldots, x_n)=1} \Sigma_{i=1}^n a_i x_i$$

where $a_1, \ldots, a_n$ are the proper values of $[A]$. The right side represents of course, $\Psi(a_1, \ldots, a_n) = \Psi(A)$.

Replacing $X$ by $\theta X$, where $\theta$ is a complex number with $|\theta| = 1$ we see that $\Phi(\theta X) = \Phi(X)$ remains unchanged, while $\mathscr{R}(\theta t(XA))$ —— as $\theta$ varies —— assumes a maximum equal to $|t(XA)|$. Thus,

$$\sup_{\Phi(X)=1} \mathscr{R}\, t(XA) = \sup_{\Phi(X)=1} |t(XA)| .$$

To complete the proof it is sufficient to point out that $t(XA) = (A, X^*)$ and

$$\Phi(X) = \Phi(X^*) .$$

The last equality holds since, $X^*X$ and $XX^*$, and therefore also $[X]$ and $[X^*]$ have the same proper values.

Remark. It is important to note that Definition 5 is still available and the above proof applies verbatim also in the absence of the triangle property for $\Phi$ ((iii) of Definition 4).

Theorem 5. *Let $\Phi(u_1, \ldots, u_n)$ be a symmetric gauge function on $\mathfrak{L}_n$. Then, $\Phi(A)$ is a unitarily invariant crossnorm on $\mathfrak{R}_n$. Moreover, every unitarily invariant crossnorm on $\mathfrak{R}_n$ is obtained in such a manner from a suitable symmetric gauge function on $\mathfrak{L}_n$. Thus, the class of unitarily invariant crossnorms on $\mathfrak{R}_n$ and the class of symmetric gauge functions on $\mathfrak{L}_n$ generate each other.*

*Proof.* Let $\Phi(u_1, \ldots, u_n)$ be given. Definition 5 determines the value $\Phi(A)$ for all $A \in \mathfrak{R}_n$. We verify that $\Phi(A)$ satisfies conditions i) — v) of Definition 1.

i): $\Phi(A) \geqq 0$ is clear. Now, $0 = \Phi(A) = \Phi(a_1, \ldots, a_n)$, implies $a_1 = \cdots = a_n = 0$, hence $A^*A = 0$ and thus $A = 0$.

ii): Since $(cA)^*(cA) = |c|^2 A^*A$, the numbers $|c|\, a_1, \ldots, |c|\, a_n$ represent the proper values of $[cA]$. Hence,

$$\Phi(cA) = \Phi(|c|\, a_1, \ldots, |c|\, a_n) = |c|\, \Phi(a_1, \ldots, a_n) = |c|\, \Phi(A) .$$

iii): We recall that any gauge function $\Phi(u_1, \ldots, u_n)$ on $\mathfrak{L}_n$ is the associate with its associate $\Psi(v_1, \ldots, v_n)$. Theorem 4 gives,

$$\Phi(A+B) = \sup_{\Psi(X)=1} |(A+B, X)| \leq$$
$$\leq \sup_{\Psi(X)=1} |(A, X)| + \sup_{\Psi(X)=1} |(B, X)| = \Phi(A) + \Phi(B) .$$

iv): We assume $A = \varphi \otimes \overline{\psi}$ with $\varphi \neq 0$ and $\psi \neq 0$. Then, $(\varphi \otimes \overline{\psi})^* (\varphi \otimes \overline{\psi}) = \|\varphi\|^2 \psi \otimes \overline{\psi}$ and therefore

$$[\varphi \otimes \overline{\psi}] = \|\varphi\| \, \|\psi\| \, \psi_1 \otimes \overline{\psi}_1$$

where $\psi_1 = \frac{\psi}{\|\psi\|}$ is of norm 1. Consequently, $\|\varphi\| \, \|\psi\|$ is the only positive proper value for $[\varphi \otimes \overline{\psi}]$ and

$$\Phi(\varphi \otimes \overline{\psi}) = \Phi(\|\varphi\| \, \|\psi\|, 0, \ldots, 0) = \|\varphi\| \, \|\psi\| = \|\varphi \otimes \overline{\psi}\| .$$

v): This is clear since —— as we have already pointed out before —— $[U A V^*]$ and $[A]$ have the same proper values.

Conversely. Assume that $\alpha$ is a given unitarily invariant crossnorm on $\mathfrak{R}_n$. Let $A \in \mathfrak{R}_n$. Observe first that a unitary $U$ can be found so that $A = U[A]$. Consequently,

$$\alpha(A) = \alpha([A]) .$$

Moreover, if $A$ and $\tilde{A}$ are two operators for which $[A]$ and $[\tilde{A}]$ have the same proper values (and multiplicities), then a unitary $V$ can be constructed so that $V[\tilde{A}]V^* = [A]$, and therefore

$$\alpha(A) = \alpha([A]) = \alpha([\tilde{A}]) = \alpha(\tilde{A}) .$$

Choose a fixed basis $\psi_1, \ldots, \psi_n$ in $\mathfrak{H}_n$. For an $n$-tuple $(u_1, \ldots, u_n)$ define

$$\Phi(u_1, \ldots, u_n) = \alpha(\Sigma_{i=1}^n u_i \, \psi_i \otimes \overline{\psi}_i) .$$

We verify without the slightest difficulty that the so defined $\Phi$ is a symmetric gauge function on $\mathfrak{L}_n$: Applying the last equation to $(u_1, \ldots, u_n)$, to $(c u_1, \ldots, c u_n)$, to $(u_1', \ldots, u_n')$ and to $(u_1 + u_1', \ldots, u_n + u_n')$ respectively, we verify properties i), ii) and iii) of Definition 4. The unitary invariance of $\alpha$ implies iv), while the cross-property of $\alpha$ implies v).

By what was already shown the gauge function $\Phi(u_1, \ldots, u_n)$ generates the crossnorm $\Phi(A)$. It remains to show that $\Phi(A) = \alpha(A)$ for all $A$ in $\mathfrak{R}_n$. This is easy: Let $A$ be given and $a_1, \ldots, a_n$ represent the proper values of $[A]$. By definition, $\Phi(A) = \Phi(a_1, \ldots, a_n)$. On the other hand, for some unitary $U$, we have

$$U[A]U^* = \Sigma_{i=1}^n a_i \, \psi_i \otimes \overline{\psi}_i$$

and therefore

$$\alpha(A) = \alpha([A]) = \alpha(\Sigma_{i=1}^n a_i \, \psi_i \otimes \overline{\psi}_i) = \Phi(a_1, \ldots, a_n) .$$

Thus, $\Phi(A) = \alpha(A)$.

Remark. Let $\Phi$ be a symmetric gauge function on $\mathfrak{L}_n$. Then $\Phi(A)$ is a unitarily invariant crossnorm on $\mathfrak{R}_n$. The interesting part is naturally the triangle inequality, that is, $\Phi(A+B) \leqq \Phi(A) + \Phi(B)$. The last is equivalent to saying that $\Phi(A) \leqq 1$ and $\Phi(B) \leqq 1$ implies $\Phi(pA + qB) \leqq 1$ whenever $p, q \geqq 0$ and $p + q = 1$. We may thus express the following: *Let $p \geqq 0$, $q \geqq 0$, $p + q = 1$, and let $a_1, \ldots, a_n$; $b_1, \ldots, b_n$ and $c_1, \ldots, c_n$ denote the proper values (each appearing the number of times equal to its mutiplicity) of $[A]$, $[B]$ and $[pA + qB]$ respectively. Then, for any symmetric gauge function $\Phi$ on $\mathfrak{L}_n$, the inequalities $\Phi(a_1, \ldots, a_n) \leqq 1$, $\Phi(b_1, \ldots, b_n) \leqq 1$, imply $\Phi(c_1, \ldots, c_n) \leqq 1$.*

Theorem 6. *The bound $\|A\|$ of an operator $A$ represents the least unitarily invariant crossnorm on $\mathfrak{R}_n$.*

*Proof.* Let $\alpha$ be a unitarily invariant crossnorm and $\Phi$ the symmetric gauge function which generates it. Representing $A$ in the polar form $\Sigma_{i=1}^{n} \lambda_i \varphi_i \otimes \overline{\psi}_i$ we have,

$$\|A\| = \max_{1 \leqq i \leqq n} \lambda_i \leqq \Phi(\lambda_1, \ldots, \lambda_n) = \alpha(A) .$$

That $\|A\|$ itself is a unitarily invariant crossnorm is clear.

Theorem 7. *Let $\alpha$ be a unitarily invariant crossnorm on $\mathfrak{R}_n$. Then, $\alpha'$ is also a unitarily invariant crossnorm. Moreover, if $\alpha$ is generated by the symmetric gauge function $\Phi$ on $\mathfrak{L}_n$, then $\alpha'$ is generated by the symmetric gauge function $\Psi$ associate with $\Phi$.*

*Proof.* For $X \in \mathfrak{R}_n$ we have $\alpha(X) = \Phi(X)$. Thus,

$$\alpha'(A) = \sup_{\alpha(X)=1} |t(XA)| = \sup_{\Phi(X)=1} |t(XA)| = \Psi(A);$$

the equality sign on the extreme right being justified by Theorem 4.

Corollary. *A unitarily invariant crossnorm $\alpha$ on $\mathfrak{R}_n$ is reflexive, that is, satisfies the condition $\alpha'' = \alpha$.*

*Proof.* Let $\Phi$ be the gauge function generating $\alpha$ and $\Psi$ denote the associate with $\Phi$. By Lemma 12, the associate with $\Psi$ is again $\Phi$. The above theorem states that $\alpha'$ is generated by $\Psi$, and similarly $\alpha''$ is generated by the associate with $\Psi$, that is, by $\Phi$. Since $\Phi$ generates both $\alpha$ and $\alpha''$, we have $\alpha = \alpha''$.

## 5. The symmetric gauge functions on $\mathfrak{L}$ and the unitarily invariant crossnorms on $\mathfrak{R}$

We are ready to extend our results to an infinite-dimensional Hilbert space $\mathfrak{H}$ and indicate (among other things) how the symmetric gauge function on the linear space $\mathfrak{L}$ of sequences of real numbers having only a finite number of non-zero terms and the unitarily invariant crossnorms on the linear space $\mathfrak{R}$ of all operators on $\mathfrak{H}$ of finite rank, generate each other.

Definition 6. Let $\mathfrak{L}$ be the set of all infinite sequences of real numbers $(u_1, u_2, \dots)$ having only a finite number of non-zero terms. With the obvious definition of addition and scalar multiplication, $\mathfrak{L}$ is a linear space. A function $\Phi = \Phi(u_1, u_2, \dots)$ on $\mathfrak{L}$ is a symmetric gauge function if it satisfies conditions i) — v) of Definition 4 when the $n$-tuples are replaced by points of $\mathfrak{L}$.

We verify without the slightest difficulty that Lemma 6 —— which has played an important part in the preceding discussion —— extends in a straightforward manner also to symmetric gauge functions on $\mathfrak{L}$.

Clearly, the set of all points $(u_1, \dots, u_n, 0, 0, \dots)$ of $\mathfrak{L}$ with $u_i = 0$ for $i > n$, may be identified with $\mathfrak{L}_n$. A given symmetric gauge function $\Phi(u_1, u_2, \dots)$ on $\mathfrak{L}$ defines a symmetric gauge function

$$\Phi_n(u_1, \dots, u_n) = \Phi(u_1, \dots, u_n, 0, 0, \dots)$$

on $\mathfrak{L}_n$ for $n = 1, 2, \dots$. Each $\Phi_n(u_1, \dots, u_n)$ determines on $\mathfrak{L}_n$ an associate $\Psi_n(v_1, \dots, v_n)$.

Lemma 13. *We have,*

$$\Psi_n(v_1, \dots, v_n) = \Psi_{n+1}(v_1, \dots, v_n, 0) = \Psi_{n+2}(v_1, \dots, v_n, 0, 0) = \dots .$$

Let $\Psi(v_1, \dots, v_n, 0, 0, \dots)$ *stand for their common value.*

*Proof.* By definition

$$\begin{aligned}\Psi_n(v_1, \dots, v_n) &= \max_{(u_1, \dots, u_n)} \frac{u_1 v_1 + \dots + u_n v_n}{\Phi_n(u_1, \dots, u_n)} = \\ &= \max_{(u_1, \dots, u_n)} \frac{u_1 v_1 + \dots + u_n v_n}{\Phi_{n+1}(u_1, \dots, u_n, 0)} \leqq \\ &\leqq \max_{(u_1, \dots, u_{n+1})} \frac{u_1 v_1 + \dots + u_n v_n + u_{n+1} 0}{\Phi_{n+1}(u_1, \dots, u_n, u_{n+1})} = \\ &= \Psi_{n+1}(v_1, \dots, v_n, 0) .\end{aligned}$$

On the other hand,

$$\Phi_n(u_1, \dots, u_n) = \Phi_{n+1}(u_1, \dots, u_n, 0) \leqq \Phi_{n+1}(u_1, \dots, u_n, u_{n+1})$$

implies

$$\frac{u_1 v_1 + \dots + u_n v_n + u_{n+1} 0}{\Phi_{n+1}(u_1, \dots, u_n, u_{n+1})} \leqq \frac{u_1 v_1 + \dots + u_n v_n}{\Phi_n(u_1, \dots, u_n)}$$

and therefore,

$$\Psi_{n+1}(v_1, \dots, v_n, 0) \leqq \Psi_n(v_1, \dots, v_n) .$$

The rest of the proof is clear.

Lemma 14. *For a given symmetric gauge function $\Phi(u_1, u_2, \dots)$ on $\mathfrak{L}$, the $\Psi(v_1, v_2, \dots)$ is also a symmetric gauge function.*

*Proof.* Clearly, $\Psi$ satisfies conditions (i), (ii), (iv) and (v). Thus it is sufficient to verify (iii). Choose two points in $\mathfrak{L}$. These may be written in the form $(v_1, \ldots, v_N, 0, 0, \ldots)$ and $(v'_1, \ldots, v'_N, 0, 0, \ldots)$. We have,

$$\begin{aligned}\Psi(v_1+v'_1, \ldots, v_N+v'_N, 0, 0, \ldots) &= \Psi_N(v_1+v'_1, \ldots, v_N+v'_N) \leqq \\ &\leqq \Psi_N(v_1, \ldots, v_N) + \Psi_N(v'_1, \ldots, v'_N) = \\ &= \Psi(v_1, \ldots, v_N, 0, 0, \ldots) + \Psi(v'_1, \ldots, v'_N, 0, 0, \ldots)\,.\end{aligned}$$

This concludes the proof.

We term $\Psi(v_1, v_2, \ldots)$ the *associate* with $\Phi(u_1, u_2, \ldots)$.

Remark. It is easy to see that for a given $(v_1, v_2, \ldots) \in \mathfrak{L}$,

$$\Psi(v_1, v_2, \ldots) = \sup_{(u_1, u_2, \ldots) \in \mathfrak{L}} \frac{u_1 v_1 + u_2 v_2 + \cdots}{\Phi(u_1, u_2, \ldots)}\,.$$

Lemma 15. *Any symmetric gauge function $\Phi(u_1, u_2, \ldots)$ on $\mathfrak{L}$ is at the same time the associate with its associate $\Psi(v_1, v_2, \ldots)$.*

*Proof.* Denote by $X$ the associate with $\Psi$. Choose a point $(u_1, \ldots u_N, 0, 0, \ldots)$ in $\mathfrak{L}$. As above, form $\Phi_N$, $\Psi_N$ and $X_N$ for $\Phi$, $\Psi$ and $X$ respectively. By Lemma 12, we have $\Phi_N = X_N$ and therefore,

$$\begin{aligned}X(u_1, \ldots, u_N, 0, 0, \ldots) &= X_N(u_1, \ldots, u_N) = \\ &= \Phi_N(u_1, \ldots, u_N) = \Phi(u_1, \ldots, u_N, 0, 0, \ldots)\,.\end{aligned}$$

Thus, $X = \Phi$ on $\mathfrak{L}$.

Theorem 8. *Let $\Phi(u_1, u_2, \ldots)$ be a symmetric gauge function on $\mathfrak{L}$. For an operator $A$ on $\mathfrak{H}$ of finite rank define*

$$\Phi(A) = \Phi(a_1, a_2, \ldots)$$

*where $a_1, a_2, \ldots$ are the proper values of $[A]$; the multiplicity of a non-zero proper value of $[A]$ equals to the number of times it appears in the above sequence. Then $\Phi(A)$ is a unitarily invariant crossnorm on $\mathfrak{R}$. Moreover, every unitarily invariant crossnorm is obtained in such a manner from a suitable symmetric gauge function. Thus, the unitarily invariant crossnorms on $\mathfrak{R}$ and the class of symmetric gauge functions on $\mathfrak{L}$ generate each other.*

*Proof.* Assume first that $\Phi(u_1, u_2, \ldots)$ is given. Clearly, $\Phi(A)$ is then defined for all $A$ in $\mathfrak{R}$. Furthermore, if $\mathscr{K}$ is any fixed finite-dimensional subspace of $\mathfrak{H}$, then $\Phi(A)$ is a norm if $A$ is restricted to those operators for which the ranges of both $A$ and $A^*$ are in $\mathscr{K}$. Now, given any two operators $\tilde{A}$, $\tilde{B}$, of finite rank, let $\mathscr{K}$ be spanned by the ranges of $\tilde{A}$, $\tilde{A}^*$, $\tilde{B}$, $\tilde{B}^*$. Clearly, $\mathscr{K}$ is finite-dimensional; $\Phi(A)$ is thus a norm for a family of operators which contains both (arbitrarily given!) operators $\tilde{A}$, $\tilde{B}$. The defining properties of a norm involve no more than two operators at a time. This implies that the defined above $\Phi$ is a norm on $\mathfrak{R}$.

Moreover, the argument used in the finite-dimensional case (Theorem 5) also proves the unitary invariance as well as the cross-property for $\Phi(A)$.

Conversely. Assume that $\mathfrak{H}$ is infinite-dimensional and $\alpha$ is a unitarily invariant crossnorm on $\mathfrak{R}$. Choose a fixed infinite orthonormal sequence of vectors $\{\psi_i\}$ in $\mathfrak{H}$. For $(u_1, u_2, \ldots) \in \mathfrak{L}$ define

$$\Phi(u_1, u_2, \ldots) = \alpha(\Sigma_i \mu_i \, \psi_i \otimes \overline{\psi}_i) .$$

The argument employed in the finite-dimensional case also applies here: $\Phi(u_1, u_2, \ldots)$ is a symmetric gauge function on $\mathfrak{L}$ and the crossnorm $\Phi(A)$ it generates on $\mathfrak{R}$, coincides with $\alpha(A)$.

Theorem 9. *Let $\Phi$ be a symmetric gauge function on $\mathfrak{L}$ and $\alpha$ the unitarily invariant crossnorm it generates on $\mathfrak{R}$. Then its associate gauge function $\Psi$ generates the crossnorm $\alpha'$ associate with $\alpha$.*

*Proof.* Theorem 7 proves this when $\mathfrak{H}$ is finite-dimensional. It remains to extend the proof to the infinite-dimensional case. So let $A \in \mathfrak{R}$. We show that

$$\alpha'(A) = \Psi(a_1, a_2, \ldots) = \Psi(A) ,$$

where $a_1, a_2, \ldots$ are the proper values (each non-zero proper value appearing the number of times equal to its multiplicity) of $[A]$.

Let $\mathfrak{M}$ be the finite say $m$-dimensional subspace of $\mathfrak{H}$ spanned by the ranges of $A$ and $A^*$. An operator on $\mathfrak{M}$ being in the form $\Sigma_{i=1}^{n} \lambda_i \varphi_i \otimes \overline{\psi}_i$ with $\lambda_i > 0$ and the $\varphi_i$'s as well as the $\psi_i$'s in $\mathfrak{M}$ defines of course, also an operator on $\mathfrak{H}$. Thus the linear space of all operators on $\mathfrak{M}$ may be identified with the linear manifold of all those operators $X$ on $\mathfrak{H}$, whose range together with the range of its adjoint $X^*$ is included in $\mathfrak{M}$.

Put $\Phi(u_1, \ldots, u_m, 0, 0, \ldots) = \Phi_m(u_1, \ldots, u_m)$ and $\Psi(u_1, \ldots, u_m, 0, 0, \ldots) = \Psi_m(u_1, \ldots, u_m)$. Then $\Phi_m$ and $\Psi_m$ are associate with each other. Moreover,

$$\Phi(X) = \Phi_m(X) \quad \text{and} \quad \Psi(X) = \Psi_m(X)$$

for all operators $X$ on $\mathfrak{M}$. The already proven finite-dimensional case implies that $\Psi_m(A)$ is the least of the constants $c$ satisfying the inequality $|t(XA)| \leqq c\, \Phi_m(X)$ for all operators $X$ on $\mathfrak{M}$, hence in the light of the identification introduced above, also the inequality $|t(XA)| \leqq c\, \Phi(X)$ for all those operators $X$ on $\mathfrak{H}$, whose range together with the range of its adjoint $X^*$ is in $\mathfrak{M}$. The supremum defining $\Psi_m(A)$ is thus not greater than the one for $\alpha'(A)$. Consequently,

$$\alpha'(A) \geqq \Psi_m(A) = \Psi(A) .$$

On the other hand, if $X \in \mathfrak{R}$ is given, choose a finite, say $(m+k)$-dimensional subspace containing the ranges of $A$, $A^*$, $X$, $X^*$. Then,

$$\frac{|t(XA)|}{\Phi(X)} = \frac{|t(XA)|}{\Phi_{m+k}(X)} \leqq \Psi_{m+k}(A) = \Psi_m(A) = \Psi(A) .$$

Thus also, $\alpha'(A) \leqq \Psi(A)$.

Theorem 10. *Every unitarily invariant crossnorm* $\alpha$ on $\mathfrak{R}$ *satisfies the condition* $\alpha'' = \alpha$.

*Proof.* Let $\Phi$ be the symmetric gauge function generated by $\alpha$. By Theorem 9, $\alpha'$ generates the associate with $\Phi$, and again $\alpha''$ generates the associate with the associate of $\Phi$, which coincides with $\Phi$ (by Lemma 15). Thus, $\alpha''$ must coincide with $\alpha$ since they generate the same $\Phi$.

Theorem 11. *A crossnorm* $\alpha$ on $\mathfrak{R}$ *is unitarily invariant if and only if it is uniform.*

*Proof.* Assume that $\alpha$ is unitarily invariant and $\Phi$ is the symmetric gauge function it generates on $\mathfrak{L}$. Let $A$ be of finite rank and $X$ represent an arbitrary operator. Let $\lambda_1 \geqq \lambda_2 \geqq \ldots$ and $\mu_1 \geqq \mu_2 \geqq \ldots$ represent the proper values of $[A]$ and $[XA]$ respectively. By Lemma I.8 we have $\mu_i \leqq \|X\| \lambda_i$. Consequently,

$$\begin{aligned}\alpha(XA) &= \Phi(XA) = \Phi(\mu_1, \mu_2, \ldots) \leqq \\ &\leqq \Phi(\|X\| \lambda_1, \|X\| \lambda_2, \ldots) = \|X\| \Phi(\lambda_1, \lambda_2, \ldots) = \\ &= \|X\| \Phi(A) = \|X\| \alpha(A) ;\end{aligned}$$

the inequality sign being justified by the extension of Lemma 6. It follows that

$$\begin{aligned}\alpha(AY) &= \alpha((AY)^*) = \alpha(Y^*A^*) \leqq \\ &\leqq \|Y^*\| \alpha(A^*) = \|Y\| \alpha(A) .\end{aligned}$$

and therefore also,

$$\alpha(XAY) \leqq \|X\| \alpha(AY) \leqq \|X\| \|Y\| \alpha(A) .$$

Thus, unitary invariance for $\alpha$ implies uniformity. The converse is easy and was stated in Lemma 1.

We summarize a part of the preceding discussion:

Theorem 12. *The class of unitarily invariant crossnorms on* $\mathfrak{R}$ *(on* $\mathfrak{R}_n$*) coincides with the class of uniform crossnorms. Either class and the class of symmetric gauge functions on* $\mathfrak{L}$ *(on* $\mathfrak{L}_n$*) generate each other.*

## 6. A special class of symmetric gauge functions

Below, we list some of the symmetric gauge functions which are frequently encountered in analysis and discuss a few of their properties. Of course, this discussion also carries over to the corresponding class of unitarily invariant crossnorms inasmuch as they generate each other. First however, we make the following notational remark: In the past $\Phi(X)$ stood for the norm generated by $\Phi(\lambda_1, \lambda_2, \ldots)$. We also find it convenient to write $\alpha(\lambda_1, \lambda_2, \ldots)$ for the symmetric gauge function generated by the unitarily invariant crossnorm $\alpha(X)$. Moreover, in the light of Theorem 9, we are also justified to write $\Phi'$ for the gauge function associate with $\Phi$.

For $(\lambda_1, \lambda_2, \ldots)$ in $\mathfrak{L}$, define

$$\Phi^{(p)}(\lambda_1, \lambda_2, \ldots) = (\Sigma_i |\lambda_i|^p)^{\frac{1}{p}} \qquad \text{if } 1 \leqq p < +\infty$$

and put

$$\Phi^{(\infty)}(\lambda_1, \lambda_2, \ldots) = \max |\lambda_i| .$$

It is not difficult to see that $\Phi^{(p)}$ is a symmetric gauge function on $\mathfrak{L}$ for $1 \leqq p \leqq +\infty$. Only condition (iii) of Definition 6 is not immediately obvious. For $1 < p < +\infty$ this condition is the well-known MINKOWSKI inequality, while the case $p = 1$ and $p = +\infty$ is easy to verify directly.

The unitarily invariant crossnorms on $\mathfrak{R}$ generated by the above gauge functions are obviously

$$\Phi^{(p)}(X) = (t([X]^p))^{\frac{1}{p}} \qquad \text{for } 1 \leqq p < +\infty$$

$$\Phi^{(\infty)}(X) = \|X\| .$$

In particular

$$\Phi^{(1)}(X) = \tau(X) ,$$

$$\Phi^{(2)}(X) = \sigma(X) .$$

Let $1 \leqq p \leqq +\infty$, and $\frac{1}{p} + \frac{1}{q} = 1$. The well-known HÖLDER inequality states that the gauge functions $\Phi^{(p)}$ and $\Phi^{(q)}$ are associate with each other. By Theorem 9, the same is also true for the corresponding unitarily invariant crossnorms.

The special significance of $\Phi^{(2)} = \sigma$ is evident from the two theorems which follow. First however, we readily verify the following simple relations valid for any two symmetric gauge functions (unitarily invariant crossnorms) $\alpha$ and $\beta$.

(1) $\alpha \leqq \beta$ implies $\alpha' \geqq \beta'$.

(2) $\frac{\alpha + \beta}{2}$ is also a symmetric gauge function (unitarily invariant crossnorm) and $\left(\frac{\alpha + \beta}{2}\right)' \leqq \frac{\alpha' + \beta'}{2}$.

Theorem 13. *We have $\sigma = \sigma'$. Moreover, $\sigma$ is the only self-associate symmetric gauge function (unitarily invariant crossnorm).*

*Proof.* It is an immediate consequence of the definition of the associate that for any two points $u$, $v$ in $\mathfrak{L}$ (operators $X$, $Y$ in $\mathfrak{R}$) we have

$$(u, v) \leqq \sigma'(v)\, \sigma(u) .$$

In particular if $u = v$,

$$(\sigma(u))^2 = (u, u) \leqq \sigma'(u)\, \sigma(u) ,$$

that is, $\sigma \leqq \sigma'$. But then also $\sigma' \geqq \sigma'' = \sigma$. Thus $\sigma = \sigma'$, that is, $\sigma$ is self-associate.

Finally, if $\alpha$ is also self-associate then

$$(\sigma(u))^2 = (u, u) \leqq \alpha'(u)\, \alpha(u) = (\alpha(u))^2$$

implies $\sigma \leqq \alpha$. Therefore also $\sigma = \sigma' \geqq \alpha' = \alpha$. Thus, $\sigma = \alpha$.

Theorem 14. *Let* $\alpha$ *be an arbitrary symmetric gauge function on* $\mathfrak{L}$ *(unitarily invariant crossnorm on* $\mathfrak{R}$*). Define*

$$\alpha_1 = \frac{\alpha + \alpha'}{2} \quad \textit{and} \quad \alpha_n = \frac{\alpha_{n-1} + (\alpha_{n-1})'}{2} \quad \textit{for} \quad n > 1 .$$

*Then the sequence* $\{\alpha_n\}$ *converges decreasingly to* $\sigma$.

*Proof.* It is a consequence of (2) above and $(\alpha_{n-1})'' = \alpha_{n-1}$ that

$$(\alpha_n)' \leqq \alpha_n \qquad \text{for all } n .$$

Hence also

$$\alpha_n \leqq \alpha_{n-1} \qquad \text{for } n > 1 .$$

It is a consequence of the last two inequalities that for every pair of natural numbers $p$ and $q$ we have, $(\alpha_p)' \leqq (\alpha_{p+q})' \leqq \alpha_{p+q} \leqq \alpha_q$. Thus,

$$(\alpha_p)' \leqq \alpha_q .$$

It follows that

$$(\alpha_1)' \leqq (\alpha_2)' \leqq (\alpha_3)' \leqq \ldots \leqq \alpha_3 \leqq \alpha_2 \leqq \alpha_1 .$$

We also have,

$$\alpha_2 - (\alpha_2)' \leqq \alpha_2 - (\alpha_1)' = \frac{\alpha_1 + (\alpha_1)'}{2} - (\alpha_1)' = \frac{\alpha_1 - (\alpha_1)'}{2} ,$$

$$\alpha_3 - (\alpha_3)' \leqq \alpha_3 - (\alpha_2)' = \frac{\alpha_2 - (\alpha_2)'}{2} \leqq \frac{\alpha_1 - (\alpha_1)'}{2^2} ,$$

and in general

$$\alpha_n - (\alpha_n)' \leqq \frac{\alpha_1 - (\alpha_1)'}{2^{n-1}} .$$

Put $\lim_n \alpha_n = \beta$. Since $\lim_n (\alpha_n - (\alpha_n)') = 0$ we have also $\lim_n (\alpha_n)' = \beta$. It is clear that $\beta$ is also a symmetric gauge function (unitarily invariant crossnorm). We shall prove that $\beta = \beta'$. Since, $\alpha_n \geqq \beta$, we have $(\alpha_n)' \leqq \beta'$ and thus $\beta = \lim_n (\alpha_n)' \leqq \beta'$. Similarly, $(\alpha_n)' \leqq \beta$, implies $\alpha_n \geqq \beta'$ and also $\beta = \lim_n \alpha_n \geqq \beta'$. Thus, $\beta = \beta'$. By Theorem 13, necessarily $\beta = \sigma$.

## 7. Norm ideals and the minimal norm ideals

Throughout the rest of this chapter we assume that $\alpha$ is a fixed crossnorm on $\mathfrak{R}$ whose associate $\alpha'$ is also a crossnorm. This, we know, is the case if and only if $\alpha(X) \geqq \|X\|$ for all $X \in \mathfrak{R}$. In particular, this is true whenever $\alpha$ is unitarily invariant.

Definition 7. Defining $\alpha$ on $\mathfrak{R}$ one obtains the normed linear space $\mathfrak{R}(\alpha)$. Its (abstract) metric completion (compare HAUSDORFF [1] p. 106) will be denoted by $\mathfrak{R}_\alpha$. [1]

[1] Considered as Banach spaces the $\mathfrak{R}_\alpha$ are the cross-spaces in the terminology of SCHATTEN [6]. There, the linear space of all operators on $\mathfrak{H}$ of finite rank, is denoted by $\mathfrak{H} \odot \bar{\mathfrak{H}}$. Defining on it a crossnorm $\alpha$ one obtains the normed linear space $\mathfrak{H} \odot_\alpha \bar{\mathfrak{H}}$. The metric completion of the last, defines the cross-space generated by $\alpha$ and is denoted by $\mathfrak{H} \otimes_\alpha \bar{\mathfrak{H}}$.

We shall see below that $\mathfrak{R}_\alpha$ may be identified with a Banach space of operators. Moreover, the $\mathfrak{R}_\alpha$ generated by means of the unitarily invariant $\alpha$, are precisely the minimal norm ideals of (necessarily completely continuous) operators.

Definition 8. For an operator $A$ on $\mathfrak{H}$ define

$$\|A\|_\alpha = \sup_{0 \neq X \in \mathfrak{R}} \frac{|t(XA)|}{\alpha(X)}.$$

If $\|A\|_\alpha < +\infty$, we term $A$ *of finite $\alpha$-norm.*

Lemma 16. *Whenever $A$ is of finite rank, then*

$$\alpha'(A) = \|A\|_\alpha < +\infty .$$

Lemma 17. *We have always*

$$\|A\| \leqq \|A\|_\alpha \leqq \tau(A) .$$

*Proof.* Clearly,

$$|(A\varphi, \psi)| = |t((\varphi \otimes \overline{\psi})A)| \leqq$$
$$\leqq \|A\|_\alpha \alpha(\varphi \otimes \overline{\psi}) = \|A\|_\alpha \|\varphi\| \, \|\psi\|$$

implies $\|A\| \leqq \|A\|_\alpha$. To prove the remaining inequality, we may assume that $\tau(A) < +\infty$, that is, $A \in (\tau c)$. But then

$$|t(XA)| \leqq \tau(A) \, \|X\| \leqq \tau(A) \, \alpha(X)$$

furnishes the desired conclusion.

Corollary. $\|A\|_\alpha = 0$ *if and only if* $A = 0$.

*Proof.* If $\|A\|_\alpha = 0$, then $\|A\| \leqq \|A\|_\alpha$ implies $A = 0$. Conversely, $A = 0$ implies $t(XA) = 0$ for all $X \in \mathfrak{R}$ and hence $\|A\|_\alpha = 0$.

Theorem 15. *Let $A$ be a fixed operator with $\|A\|_\alpha < +\infty$. As $X$ varies in $\mathfrak{R}(\alpha)$, the expression*

$$t(XA)$$

*represents a linear functional with a bound equal to $\|A\|_\alpha$. Moreover, every bounded linear functional on $\mathfrak{R}(\alpha)$ is obtained in such a manner (from a unique $A$ with $\|A\|_\alpha < +\infty$). The above correspondence between the operators of finite $\alpha$-norm and the linear bounded functionals on $\mathfrak{R}(\alpha)$ is obviously one-to-one and linear. Consequently, the set of all operators $A$ of finite $\alpha$-norm is a linear space and $\|A\|_\alpha$ defines there a norm. The resulting normed linear space is equivalent to $(\mathfrak{R}(\alpha))^* = (\mathfrak{R}_\alpha)^*$, hence necessarily complete.*

*Proof.* Let $\mathcal{F}$ be a bounded linear functional on $\mathfrak{R}(\alpha)$. As in the proof of Theorem IV.1, F. Riesz' representation theorem for bounded linear functionals determines a unique operator $A$ such that

$$\mathcal{F}(X) = t(XA)$$

for all operators $X$ of finite rank. We have,

$$\|\mathcal{F}\| = \sup_{0 \neq X \in \mathfrak{R}} \frac{|\mathcal{F}(X)|}{\alpha(X)} = \sup_{0 \neq X \in \mathfrak{R}} \frac{|t(XA)|}{\alpha(X)} = \|A\|_\alpha .$$

Conversely. The last equality shows that whenever $\|A\|_\alpha < +\infty$, the obviously linear functional $t(XA)$ on $\mathfrak{R}(\alpha)$ is bounded and its bound is $\|A\|_\alpha$. The rest is clear.

Corollary. *$\mathfrak{R}_{\alpha'}$ is the subspace of $(\mathfrak{R}_\alpha)^*$ consisting of all those operators which are approximable there by operators of finite rank.*

*Proof.* For an operator $X \in \mathfrak{R}$, we have $\alpha'(X) = \|X\|_\alpha$. Thus,

$$\mathfrak{R}(\alpha') \subset (\mathfrak{R}(\alpha))^* = (\mathfrak{R}_\alpha)^* .$$

The right side is a complete space and may —— by the preceding theorem —— be identified with a space of operators. Hence the metric (CANTOR-MERAY) completion $\mathfrak{R}_{\alpha'}$ of $\mathfrak{R}(\alpha')$ may be identified with a subspace of operators.

Remark. We emphasize that since $\|A\|_\alpha = 0$ if and only if $A = 0$, two different elements of $\mathfrak{R}_{\alpha'}$ determine two different operators.

Theorem 16. *The Banach space $(\mathfrak{R}_\alpha)^* = (\mathfrak{R}(\alpha))^*$ is a norm ideal if and only if $\alpha$ is unitarily invariant.*

*Proof.* Theorem 11 states that $\alpha$ is unitarily invariant if and only if it is uniform.

Assume that $(\mathfrak{R}(\alpha))^*$ is a norm ideal. Let $X_0$, $Y_1$, $Y_2$ be three fixed operators where the first is of finite rank. Choose a linear functional $\mathcal{F}$ on $\mathfrak{R}(\alpha)$ with a bound 1, assuming for the operator $Y_1X_0Y_2$ of finite rank, the value $\alpha(Y_1X_0Y_2)$ (see BANACH [1, p. 55]). By Theorem 15 this functional determines an operator $A$ such that $\|A\|_\alpha = 1$ and $\mathcal{F}(X) = t(XA)$ for all $X \in \mathfrak{R}$. Hence,

$$\begin{aligned} \alpha(Y_1X_0Y_2) &= \mathcal{F}(Y_1X_0Y_2) = t(Y_1X_0Y_2A) = \\ &= t(X_0Y_2AY_1) \leqq \|Y_2AY_1\|_\alpha \, \alpha(X_0) \leqq \\ &\leqq \|A\|_\alpha \|Y_1\| \, \|Y_2\| \, \alpha(X_0) = \|Y_1\| \, \|Y_2\| \, \alpha(X_0) . \end{aligned}$$

Conversely. Assume that $\alpha$ is uniform. For every $X \in \mathfrak{R}$ we have:

$$\begin{aligned} |t(XY_1AY_2)| &= |t(Y_2XY_1A)| \leqq \\ &\leqq \|A\|_\alpha \, \alpha(Y_2XY_1) \leqq \|A\|_\alpha \|Y_1\| \, \|Y_2\| \, \alpha(X) . \end{aligned}$$

This implies

$$\|Y_1AY_2\|_\alpha \leqq \|Y_1\| \, \|Y_2\| \, \|A\|_\alpha .$$

Corollary. *Whenever $\alpha$ is unitarily invariant, then $(\mathfrak{R}_\alpha)^*$ is a Banach algebra.*

*Proof.* Clearly, $\|I\| = 1$ and $\|B\| \leqq \|B\|_\alpha$. Thus,

$$\|AB\|_\alpha = \|IAB\|_\alpha \leqq \|I\|\,\|B\|\,\|A\|_\alpha \leqq \|A\|_\alpha \|B\|_\alpha .$$

Theorem 17. *$\mathfrak{R}_\alpha$ is a minimal norm ideal if and only if $\alpha$ is unitarily invariant.*

*Proof.* Assume that $\alpha$ is unitarily invariant (uniform). Theorem 10 implies that $\alpha'' = \alpha$. Consequently,

$$\mathfrak{R}_\alpha = \mathfrak{R}_{\alpha''} \subset (\mathfrak{R}_{\alpha'})^* ,$$

and thus (not only $\mathfrak{R}_{\alpha'}$ but also) $\mathfrak{R}_\alpha$ is a Banach space of operators.

Let $X$, $Y$ be in $\mathfrak{A}$. If $\{A_n\}$ is a Cauchy sequence in $\mathfrak{R}(\alpha)$, the same may be said for $\{XA_nY\}$ since

$$\alpha(XA_nY - XA_mY) = \alpha(X(A_n - A_m)Y) \leqq$$
$$\leqq \|X\|\,\|Y\|\,\alpha(A_n - A_m) .$$

The sequences $\{A_n\}$ and $\{XA_nY\}$ determine in $\mathfrak{R}_\alpha$ the operators $A$ and $B$ respectively. Moreover, since

$$\alpha(A - A_n) \to 0 \text{ and } \alpha(B - XA_nY) \to 0 ,$$

we also have

$$\|A - A_n\| \to 0 \text{ and } \|B - XA_nY\| \to 0 .$$

Consequently,

$$\|B - XAY\| \leqq \|B - XA_nY\| + \|XA_nY - XAY\| \leqq$$
$$\leqq \|B - XA_nY\| + \|X\|\,\|Y\|\,\|A_n - A\| \to 0 ,$$

that is, $XAY = B \in \mathfrak{R}_\alpha$ and

$$\alpha(XAY) = \lim_n \alpha(XA_nY) \leqq$$
$$\leqq \|X\|\,\|Y\| \lim_n \alpha(A_n) = \|X\|\,\|Y\|\,\alpha(A) .$$

Thus, $\mathfrak{R}_\alpha$ is a norm ideal. Since every ideal must include $\mathfrak{R}$, it follows that $\mathfrak{R}_\alpha$ is the minimal norm ideal determined by $\alpha$.

Conversely. It is clear that whenever $\mathfrak{R}_\alpha$ is a norm ideal it is necessarily minimal and $\alpha$ is unitarily invariant.

Corollary. *Every minimal norm ideal $\mathfrak{J}$ is an $\mathfrak{R}_\alpha$ (for some unitarily invariant $\alpha$).*

*Proof.* The definition of a norm ideal $\mathfrak{J}$ implies that its crossnorm $\alpha$ is uniform. Clearly, $\mathfrak{R} \subset \mathfrak{J}$ and $\alpha$ is also a uniform (unitarily invariant) crossnorm on $\mathfrak{R}$. Being complete $\mathfrak{J}$ must include the minimal norm ideal $\mathfrak{R}_\alpha$. Thus, $\mathfrak{J}$ is minimal if and only if $\mathfrak{J} = \mathfrak{R}_\alpha$.

We recall that the symbol $t(Y)$ was defined for all operators $Y$ in $(\tau c)$. The last consists of all operators of the form $Y = XA$ where $X$ and

$A$ are in $(\sigma c)$; the value $t(Y)$ being independent on the particular decomposition of $Y$ as a product of two operators in $(\sigma c)$. We also have

$$|t(XA)| \leqq \sigma(A)\,\sigma(X)\,.$$

Clearly, $\sigma(X)$ is a unitarily invariant crossnorm on $\mathfrak{R}$ and

$$(\sigma c) = \mathfrak{R}_\sigma = (\mathfrak{R}_\sigma)^*\,.$$

We may say therefore that $t(XA)$ was defined for all pairs of operators $A$, $X$, for which $\|A\|_\sigma < +\infty$ and $X \in \mathfrak{R}_\sigma$. Moreover,

$$|t(XA)| \leqq \|A\|_\sigma\,\sigma(X)\,.$$

We also recall that for $X \in (\tau c)$, the symbol $t(XA)$ is well defined for every operator $A$. Again this statement may be expressed in an equivalent form which is analogous to the one stated above. Observe first that $\tau(X)$ is a unitarily invariant crossnorm on $\mathfrak{R}$ and $(\tau c) = \mathfrak{R}_\tau$. Moreover, the conjugate space of the trace-class coincides with the space of all operators, that is, every operator $A$ is finite $\tau$-norm and $\|A\|_\tau = \|A\|$. Using the last terminology, we may say that $t(XA)$ is defined for all pairs of operators $A$ and $X$ where $\|A\|_\tau < +\infty$ and $X \in \mathfrak{R}_\tau$. Moreover,

$$|t(XA)| \leqq \|A\|_\tau\,\tau(X)\,.$$

The above statements for $\sigma$ and $\tau$ motivate the stated below immediate extension for any unitarily invariant $\alpha$.

Lemma 18. *Let* $\|A\|_\alpha < +\infty$ *and* $X \in \mathfrak{R}_\alpha$. *Choose any sequence* $\{X_n\}$ *in* $\mathfrak{R}$ *such that* $\alpha(X - X_n) \to 0$. *Then* $\{t(X_n A)\}$ *converges; its limit which is thus independent on the chosen sequence* $\{X_n\}$ *is denoted by* $t(XA)$. *We also have,*

$$|t(XA)| \leqq \|A\|_\alpha\,\alpha(X)\,.$$

*Proof.* Clearly $t(XA)$ is defined for all $X \in \mathfrak{R}$. By Theorem 15 it represents a bounded linear functional $\mathscr{F}(X)$ on $\mathfrak{R}(\alpha)$. It may thus be extended to a bounded linear functional $\mathscr{F}$ on $\mathfrak{R}_\alpha$ with the same bound. This means: If $X_n \in \mathfrak{R}$ and $\alpha(X - X_n) \to 0$, then $\{t(AX_n)\} = \{\mathscr{F}(X_n)\}$ converges, namely to $\mathscr{F}(X)$.

Corollary. *If* $X \in \mathfrak{R}_\alpha$ *and* $Y \in \mathfrak{R}_{\alpha'}$, *then* $t(XY)$ *is defined.*

Lemma 19. *Whenever* $\mathfrak{R}_\alpha$ *is reflexive, then* $(\mathfrak{R}_\alpha)^*$ *is minimal.*

*Proof.* Assume that $\mathfrak{R}_\alpha$ is reflexive and $(\mathfrak{R}_\alpha)^*$ is not minimal, that is, $(\mathfrak{R}_\alpha)^*$ contains $\mathfrak{R}_{\alpha'}$ as a proper subspace. Let $B_0$ denote an operator in $(\mathfrak{R}_\alpha)^*$, (that is, $\|B_0\|_\alpha < +\infty$), which does not belong to $\mathfrak{R}_{\alpha'}$. We can find then a bounded linear functional $\mathscr{F}$ on $(\mathfrak{R}_\alpha)^*$ such that $\mathscr{F}(B_0) = 1$ and $\mathscr{F}(B) = 0$ for all $B \in \mathfrak{R}_{\alpha'}$ (see Banach [1, p. 57]). Since $\mathfrak{R}_\alpha$ is reflexive, $\mathscr{F}$ determines an operator $A \in \mathfrak{R}_\alpha$ such that $t(B_0 A) = 1$ and $t(BA) = 0$ for all $B \in \mathfrak{R}_{\alpha'}$ (recall Theorem 15). Since the operators of the form

$\varphi \otimes \overline{\psi}$ are in $\mathfrak{R}_{\alpha'}$ we shall have in particular $(A\varphi, \psi) = 0$ for all pairs of vectors $\varphi$ and $\psi$. Thus $A = 0$, in contradiction with $t(B_0 A) = 1$.

Theorem 18. *Let $\alpha$ be unitarily invariant. The following statements are equivalent:*

i) *$\mathfrak{R}_\alpha$ is reflexive,*
ii) *$\mathfrak{R}_{\alpha'}$ is reflexive,*
iii) *Both, $(\mathfrak{R}_\alpha)^*$ and $(\mathfrak{R}_{\alpha'})^*$ are minimal.*

*Proof.* We recall that a Banach space is reflexive if and only if its conjugate space is reflexive. Moreover, a subspace of a reflexive Banach space is also reflexive. In particular if $\mathfrak{R}_\alpha$ is reflexive the same is true for $(\mathfrak{R}_\alpha)^*$ and hence also for $\mathfrak{R}_{\alpha'} \subset (\mathfrak{R}_\alpha)^*$. Since $\alpha'' = \alpha$, we conclude that statements i) and ii) are equivalent. Lemma 19 proves that i) and ii) — hence also i) or ii) separately — implies iii). Finally, if iii) holds then $(\mathfrak{R}_\alpha)^* = \mathfrak{R}_{\alpha'}$ and $(\mathfrak{R}_{\alpha'})^* = \mathfrak{R}_{\alpha''} = \mathfrak{R}_\alpha$. Thus, $\mathfrak{R}_\alpha$ is reflexive.

## 8. The uniqueness of norm for the minimal norm ideals

The preceding discussion shows that every unitarily invariant crossnorm $\alpha$ on $\mathfrak{R}$ determines a unique minimal norm ideal $\mathfrak{R}_\alpha$ of (necessarily completely continuous) operators. Ignoring the norm on $\mathfrak{R}_\alpha$ we may consider it as an ideal in the algebra $\mathfrak{A}$. Accordingly we shall speak below of the "norm ideal $\mathfrak{R}_\alpha$" as well as the "ideal $\mathfrak{R}_\alpha$".

It is interesting to observe that not only does $\alpha$ determine a unique ideal $\mathfrak{R}_\alpha$ in $\mathfrak{A}$, but also conversely, the ideal $\mathfrak{R}_\alpha$ uniquely determines $\alpha$. The last is to be interpreted in the following sense: Given an ideal $\mathfrak{J} \subset \mathfrak{A}$, then there is at most one $\alpha$ for which $\mathfrak{J} = \mathfrak{R}_\alpha$. We are thus concerned with a problem involving the uniqueness of norm.

Investigations in this direction carried out by EIDELHEIT [1], GELFAND and NEUMARK [1], KAPLANSKY [1] and RICKART [1], resulted in a variety of conditions under which an abstract algebra $\mathscr{A}$ admits essentially no more than one norm, that is, when any two norms $\|f\|$ and $\|f\|_1$ under which $\mathscr{A}$ is a Banach algebra are necessarily equivalent (we mean hereby, $\|f_n\| \to 0$ if and only if $\|f_n\|_1 \to 0$). For our immediate purpose it is not necessary to state these results in the generality in which they were obtained. We merely quote below, two lemmas which will suit our needs. Both may be found in RICKART [1].

We recall that a complex algebra $\mathscr{A}$ of linear transformations on a complex vector space $\mathscr{V}$ is termed $n$-fold transitive if, given $2n$ arbitrary vectors $f_1, \ldots, f_n$ and $g_1, \ldots, g_n$ in $\mathscr{V}$ with $f_1, \ldots, f_n$ linearly independent, then there is always an $A \in \mathscr{A}$ such that $Af_i = g_i$ for $i = 1, \ldots, n$. If $\mathscr{A}$ is $n$-fold transitive for $n = 1, 2, \ldots$ then $\mathscr{A}$ is termed dense. It can be shown that 2-fold transitivity implies density (See JACOBSON [1] p. 229).

Moreover, every 1-fold transitive complex Banach algebra of linear transformations on a complex vector space is necessarily dense (see for instance, RICKART [1] p. 623).

Lemma 20. *Let $\mathscr{A}$ be a complex algebra of linear transformations on a complex vector space. If $\mathscr{A}$ is 1-fold transitive and contains non-zero linear transformations of finite rank, then there is essentially at most one norm with respect to which $\mathscr{A}$ is a Banach algebra.*

Corollary. *The algebra of all operators on a Banach space admits the bound as essentially the only norm under which it is a Banach algebra* (EIDELHEIT [1]).

The second proposition we wish to quote concerns an algebra with an involution. An algebra $\mathscr{A}$ with an involution $f \to f^*$ is said to admit an *auxiliary norm*, if on it a norm $|f|$ can be defined which in addition to the multiplicative property $|fg| \leqq |f|\,|g|$ also satisfies the condition $|f|^2 = |f^* f|$. We do not require completeness of the metric space so generated by $|f|$.

Lemma 21. *Let $\mathscr{A}$ be a *-algebra admitting an auxiliary norm. Then there is at most one norm (in general different from the auxiliary norm) with respect to which $\mathscr{A}$ is a Banach algebra, that is, any two norms with respect to which $\mathscr{A}$ is a Banach algebra are necessarily equivalent.*

Corollary. *An algebra of operators on a Hilbert space closed under the operator adjoint admits essentially no more than one norm which makes it into a Banach algebra.*

Either of the above two lemmas yields the following:

Theorem 19. *For a unitarily invariant crossnorm $\alpha$ on $\mathfrak{R}$, the norm ideals $\mathfrak{R}_\alpha$ and $(\mathfrak{R}_\alpha)^*$ admit a unique norm. This means: A norm with respect to which the ideal $\mathfrak{R}_\alpha$ is a Banach algebra is equivalent to $\alpha$; similarly a norm which makes the ideal $(\mathfrak{R}_\alpha)^*$ into a Banach algebra is equivalent to $\|A\|_\alpha$.*

## 9. An open problem

Let $\alpha$ be unitarily invariant. Then $(\mathfrak{R}_\alpha)^*$ and $\mathfrak{R}_{\alpha'}$ are the maximal and the minimal norm ideal of operators of finite $\alpha$-norm. Thus,

$$(\mathfrak{R}_\alpha)^* \supset \mathfrak{R}_{\alpha'}\,.$$

The last inclusion becomes an equality if and only if $(\mathfrak{R}_\alpha)^*$ is also minimal.

This points to *the desirability of a "simple and direct" characterization of all those unitarily invariant crossnorms (or generating them symmetric gauge functions) for which the conjugate space of the minimal norm ideal generated by such a crossnorm is also minimal, that is,* $(\mathfrak{R}_\alpha)^* = \mathfrak{R}_{\alpha'}$. It is true that one may impose on $\alpha$ a variety of sufficient conditions guaranteeing that $(\mathfrak{R}_\alpha)^*$ is minimal (compare [6, pp. 124—128] by this author). However, some simple and direct conditions which are both, necessary and sufficient, are as yet unknown.

# Bibliography

ARONSZAJN, N., and K. T. SMITH: [1] Invariant subspaces of completely continuous operators. Ann. of Math. (2) **60**, **345—350** (1954).

BANACH, S.: [1] Théorie des opérations linéaires. Monografje Matematyczne. Warsaw 1932.

BONNESEN, T., and W. FENCHEL: [1] Theorie der konvexen Körper. Ergebnisse der Math. und ihrer Grenzgebiete, III 1. Berlin: Springer 1934.

CALKIN, J. W.: [1] Abstract symmetric boundary conditions. Trans. Amer. Math. Soc. **45**, 369—442 (1939). — [2] Two sided ideals and congruences in the ring of bounded operators in Hilbert space. Ann. of Math. (2) **42**, 839—873 (1941).

COURANT, R.: [1] Über die Eigenwerte bei den Differentialgleichungen der Mathematischen Physik. Math. Z. **7**, 1—57 (1920). — [2] Zur Theorie der linearen Integralgleichungen. Math. Ann. **89**, 161—178 (1923).

DIXMIER, J.: [1] Les fonctionnelles linéaires sur l'ensemble des opérateurs bornés d'un espace de Hilbert. Ann. of Math. (2) **51**, 387—408 (1950).

DUNFORD, N., and J. SCHWARTZ: [1] Linear operators, Part I. New York; Interscience 1958.

EIDELHEIT, M.: [1] On isomorphisms of rings of linear operators. Studia math. **9**, 97—105 (1940).

FISCHER, E.: [1] Über quadratische Formen mit reellen Koeffizienten. Mh. Math. Phys. **16**, 234—249 (1905).

FUKAMIYA, M.: [1] On a theorem of GELFAND and NEUMARK and the $B^*$-algebra. Kumamoto J. Sci., Ser. A **1**, no. 1, 17—22 (1952).

GELFAND, I. M., and M. A. NEUMARK: [1] On the imbedding of normed rings into the ring of operators in Hilbert space. Mat. Sborn. N. S. **12** (54), 197—213 (1943).

GROTHENDIECK, A.: [1] Produits tensoriels topologiques et espaces nucléaires. Memoirs Amer. Math. Soc. no. 16 (1955).

HALMOS, P. R.: [1] Measure Theory. New York: D. van Nostrand 1950. — [2] Introduction to Hilbert space and the theory of spectral multiplicity. New York: Chelsea 1951. — [3] Finite-dimensional vector spaces. Ann. of Math. Stud. no. 7. Princeton University Press 1942 or D. van Nostrand 1958.

HAMBURGER, H. L., and M. E. GRIMSHAW: [1] Linear transformations in $n$-dimensional vector space. Cambridge Univ. Press 1951.

HAUSDORFF, F.: [1] Mengenlehre (dritte Auflage). Berlin and Leipzig: W. de Gruyter 1935. Reprinted by Dover Publications, New York.

HELLINGER, E., and O. TOEPLITZ: [1] Integralgleichungen und Gleichungen mit unendlichvielen Unbekannten. Encyklopädie der Mathematischen Wissenschaften II C. 13, Leipzig 1928. Reprinted by Chelsea, New York 1952.

HILBERT, D.: [1] Grundzüge einer allgemeinen Theorie der linearen Integralgleichungen. Leipzig: Teubner 1912. Reprinted by Chelsea, New York 1952.

HILLE, E., and R. S. PHILLIPS: [1] Functional analysis and semi-groups. Amer. Math. Soc. Colloq. Publ. vol. 31, New York 1957.

JACOBSON, N.: [1] Structure theory of simple rings without finiteness assumption. Trans. Amer. Math. Soc. **57**, 228—245 (1945).

JAMES, R. C.: [1] A non-reflexive Banach space isometric with its second conjugate space. Proc. Nat. Acad. Sci. (U.S.A) **37**, 174—177 (1951).

KAPLANSKY, I.: [1] Normed algebras. Duke math. J. **16**, 399—417 (1949).

KOLMOGOROV, A. N., and S. V. FORMIN: [1] Elements of the theory of functions and functional analysis. Translated by L. F. Boron, Graylock Press, Rochester, N.Y. 1957.

KREIN, M., and D. MILMAN: [1] On extreme points of regularly convex sets. Studia math. **9**, 133—138 (1940).

LOOMIS, L. H. [1] An introduction to abstract harmonic analysis. New York: D. van Nostrand Co. **1953**.

LÖWIG, H.: [1] Komplexe euklidische Räume von beliebiger endlicher oder unendlicher Dimensionszahl. Acta sci. math. Szeged **7**, 1—33 (1934).

NEUMANN, J. von: [1] Mathematische Begründung der Quantenmechanic. Nachr. Ges. Wiss. Göttingen, Math.-Phys. Kl. **1927**, 1—57. — [2] Allgemeine Eigenwerttheorie Hermitescher Funktionaloperatoren. Math. Ann. **102**, 49—131 (1929). — [3] Über adjungierte Funktionaloperatoren. Ann. of Math. **33**, 294—310 (1932). —[4] Mathematische Grundlagen der Quantenmechanic. Die Grundlehren der Mathematischen Wissenschaften, Berlin, 1932, or Dover Publications, New York 1943. — [5] On a certain topology for rings of operators. Ann. of Math. **37**, 111—115 (1936). — [6] Some matrix-inequalities and metrization of matricspace. Tomsk Univ. Rev. **1**, 286—300 (1937). — [7] Zur Algebra der Funktionaloperationen und Theorie der normalen Operatoren. Math. Ann. **102**, 370—427 (1929—1930).

PETTIS, B. J.: [1] A note on regular Banach spaces. Bull. Amer. Math. Soc. **44**, 420—428 (1938).

RICKART, C. E.: [1] The uniqueness of norm problem in Banach algebras. Ann. of Math. (2) **51**, 615—628 (1950). — [2] Banach algebras with an adjoint operation. Ann. of Math. (2) **47**, 528—550 (1946).

RIESZ, F.: [1] Über lineare Funktionalgleichungen. Acta math. **41**, 71—98 (1918).

—, and B. SZ.-NAGY: [1] Lecons d'analyse fonctionnelle. Akademiai Kiado, Budapest, 1952, or Functional analysis (translated by F. Boron) Ungar Publ. Co., New York **1955**.

RUSTON, A. F.: [1] On the Fredholm theory of integral equations for operators belonging to the trace class of a general Banach space. Proc. London Math. Soc. (2) **53**, 109—124 (1951). — [2] Direct products of Banach spaces and linear functional equations. Proc. London Math. Soc. (3) **1**, 327—384 (1951).

SCHATTEN, R.: [1] On the direct product of Banach spaces. Trans. Amer. Math. Soc. **53**, 195—217 (1943). — [2] On reflexive norms for the direct product. Trans. Amer. Math. Soc. **54**, 498—506 (1943). — [3] The cross-space of linear transformations. Ann. of Math. **47**, 73—84 (1946). — [4] On projections with bound 1. Ann. of Math. **48**, 321—325 (1947). — [5] "Closing-up" of sequence spaces. Amer. Math. Monthly **57**, 603—616 (1950). — [6] A theory of cross-spaces. Ann. of Math. Stud. no. 26, Princeton Univ. Press 1950. — [7] The space of completely continuous operators on a Hilbert space. Math. Ann. **134**, 47—49 (1957).

—, and J. VON NEUMANN: [1] The cross-space of linear transformations II. Ann. of Math. **47**, 608—630 (1946). — [2] The cross-space of linear transformations III. Ann. of Math. **49**, 557—582 (1948).

SCHMIDT, E.: [1] Über die Auflösung linearer Gleichungen mit unendlich vielen Unbekannten. Rend. Circolo Mat. di Palermo **25**, 53—77 (1908). — [2] Auflösung der allgemeinen linearen Integralgleichung. Math. Ann. **64**, 161—174 (1907). — [3] Entwicklung willkürlicher Funktionen nach Systemen vorgeschriebener. Math. Ann. **63**, 433—476 (1907).

SEGAL, I. E.: [1] Two sided ideals in operator algebras. Ann. of Math. **50**, 856—865 (1949).

SHERMAN, S.: [1] The second adjoint of a ($C^*$)-algebra. Proc. Internat. Congress Math. **1**, 470 (1950).

STONE, M. H.: [1] Linear transformations in Hilbert space and their applications to analysis. Amer. Math. Soc. Colloq. Publ. **15**, New York, 1932.

SZ.-NAGY, B. VON: [1] Spektraldarstellung linearer Transformationen des Hilbertschen Raumes. Ergebnisse der Math. Berlin: J. Springer **1942**. Reprinted by Edwards Bros., Ann. Arbor, Mich. **1947**.

TAKEDA, Z.: [1] Conjugate spaces of operator algebras. Jap. Acad. **30**, 90—95 (1954).